T0324818

DEPENDABILITY OF CRITICAL COMPUTER SYSTEMS
3

Techniques Directory

DEPENDABILITY OF CRITICAL COMPUTER SYSTEMS 3

Techniques Directory

Guidelines produced by
The European Workshop on Industrial Computer Systems
Technical Committee 7 (EWICS TC7)

Edited by

P. G. BISHOP

National Power, Leatherhead, UK

ELSEVIER APPLIED SCIENCE
LONDON and NEW YORK

ELSEVIER SCIENCE PUBLISHERS LTD
Crown House, Linton Road, Barking, Essex IG11 8JU, England

Sole distributor in the USA and Canada
ELSEVIER SCIENCE PUBLISHING CO., INC.
655 Avenue of the Americas, New York, NY 10010, USA

British Library Cataloguing in Publication Data

Dependability of critical computer systems: guidelines.
3, Techniques directory
1. Computer systems. Design
I. Bishop, P. G. II. European Workshop on Industrial
Computer Systems *Technical Committee 7*
004.21

Library of Congress Cataloging-in-Publication Data
(Revised for vol. 3)

Dependability of critical computer systems.

Vol. 3 edited by P. G. Bishop.
Includes bibliographies and indexes.
Contents: 1–2. [without special title]—
3. Techniques directory.
1. Electronic digital computers—Reliability—
Congresses. I. Redmill, Felix. II. Bishop, P. G.
III. European Workshop on Industrial Computer Systems.
TC7—Systems Reliability, Safety, and Security.
QA76.9.E94D47 1988   004   88-9705

ISBN 1-85166-544-7

Printed in Northern Ireland by The Universities Press (Belfast) Ltd.

Foreword

MICHAEL HARDY

Director, General Affairs,
Directorate-General for Telecommunications, Information
Industries and Innovation,

Commission of the European Communities

The aim of the European Community's standardization policy is to ensure the harmonized implementation of international standards throughout the Community. The development of such standards by the European standards organizations, CEN, CENELEC and ETSI contributes to meeting the challenge of 1992 – the removal of technical barriers to trade and the creation of a dynamic, single market.

The implementation of harmonized, agreed standards has thus a vital part to play in the growth of the Community's information technology and communications industry. All sectors of the European economy are expected to benefit from the use of common standards. The manufacturing sector in particular, will be substantially affected in its working methods and productivity levels as information technology spreads throughout the sector. Critical real-time I.T. applications are becoming important on the shop floor and in process control in such areas as railway signalling, nuclear plant protection and hazardous chemical processing. CEN/CENELEC have established an expert group – IT Ad hoc Expert Group for Manufacturing Standards (ITAEGM) – to assess standardization needs in the fields of integrated manufacturing and process control. The work programme for the development of such standards is set out in the CEN/CENELEC Memorandum M-IT-04 of January 1990.

The European Workshop on Industrial Computing Systems (EWICS) has, through its Technical Committee 7, brought together leading figures in the area of critical computing applications and provides a centre of expertise in Europe in this domain. The Commission has sponsored the work of EWICS TC7 in producing guidance for critical applications by providing financial support for its meetings. The work of the Committee was originally presented as a set of position papers covering a range of topics from the specification to the maintenance and modification of critical computer systems. These position papers have been widely published in technical journals and company publications and have been influential in the formulation of a number of software standards.

The first two books in the series presented eleven EWICS TC7 guidelines in a revised and updated form. This third book is a directory of methods and techniques for critical computer systems and is designed to be a companion volume for the first two books, although it can also be used as a separate source of reference.

As the record shows, EWICS TC7 has made a substantial contribution and I am confident that the material presented in this directory will be a significant element in the production of dependable computing systems for critical applications. Through close and fruitful collaboration the European standards bodies and ITAEGM, the work of EWICS TC7 can be further enhanced within the European standardization process.

Preface

The aim of the EWICS TC7 committee is to enhance safety and reliability of industrial computer systems by:–

- Involvement in pre-standardization activities. EWICS TC7 guideline material has been used in IEC 880, and is being incorporated into standards produced by IEC 65A.
- Stimulating research by organising occasional seminars and a regular workshop, SAFECOMP. Since the first workshop in 1979, SAFECOMP has become widely known and respected and SAFECOMP is now held annually.
- Promoting awareness of existing good practice through the production of guidelines.

To promote awareness to wider audience, EWICS TC7 decided that the guidelines should be published as a set of volumes under the series title *Dependability of Critical Computing Systems*.

The original techniques directory guideline was produced to provide supporting material for two other guidelines; *Design for Computer Systems Safety* and *Assessment of the Safety and Reliability of Critical Computer Systems* which appear in Volume 2 of this series. In order to make the directory a more useful standalone source of reference, some additional material on software implementation and testing has been incorporated into the book to broaden the scope.

The editor now appreciates how Samuel Johnson must have felt when he was constructing his dictionary – the process is never-ending. A more comprehensive document is always possible, given more time. I hope that the material provided proves useful. I apologize in advance for any omissions that may exist in the document.

P.G. Bishop

Acknowledgements

The editor wishes to thank all EWICS TC7 contributors for their hard work in producing this techniques directory. Acknowledgements are also made to their companies for supporting this work and making their experts available. A list of the contributing members and companies is given in Appendix 1.

Grateful acknowledgement is also due to DG XIII of the Commission of the European Communities for their sponsorship of this work.

Contents

CHAPTER 1   Introduction

CHAPTER 2   Safety Analysis

CHAPTER 3   Fault Avoidance

CHAPTER 4   Fault Detection

CHAPTER 5   Failure Detection

CHAPTER 6 Failure Containment

Chapter 1

Introduction

In industry and other business activities, the use of computers as programmable electronic systems has grown considerably in the last decade. Today, many daily activities are in one way or another controlled or protected by computers. In the process industries, nuclear power generation, flexible automated manufacturing, transportation systems, telecommunication systems and banking, much may be at stake if a critical computer should fail.

One example of a serious failure in a critical computing application occurred in 1980 at the US Strategic Air Command (SAC) where the computer system gave an incorrect warning of a Soviet ballistic missile attack. The cause was traced back to a computer fault which corrupted test communications messages into attack messages.

This incident could have been prevented quite simply by applying standard error detection techniques to the communications messages. It is therefore important that the correct design and development approaches are used in producing such systems. We hope that this book and the associated volumes in the series *Dependability of Critical Computer Systems* will contribute to the safety of critical computer systems by encouraging the use of established methods for controlling the risks of using computer-based systems.

1

1.1   Scope and Structure

This volume was produced as a companion to the first two books in the series which give general guidance on a range of topics from the initial specification to maintenance and modification. This third volume is a techniques directory which is designed to be a source of reference on the specific techniques and methods used in critical applications. It should be useful as an initial point of reference for specialist software developers and assessors, and it should also be helpful to those who have a general interest in the critical software field.

The techniques directory is also intended to serve as a standalone source of reference. For this reason, the first chapter introduces the general systems approach to risk control in plant and computer systems, together with the general strategies for controlling risk. This general strategy is used as the broad classification scheme for the methods and techniques which can be applied at the computer-based system level.

The remaining chapters of the book are presented as a set of structured references grouped under the different headings associated with the main principles of risk control. While the references do not prescribe the use of particular techniques, assessments of the suitability of particular approaches are given which are based on the practical experience of EWICS TC7 members.

1.2   Critical Systems Overview

1.2.1   Definition of Terms

The following definitions will be used in this directory:–

Plant   the part of the real world to be monitored or controlled.

Target System	the system to be implemented, (normally a computer-based control system).
Hazard	a source of energy, or combination of factors that can lead to an accident if inadequately controlled.
Fault	a defect in a system which may, under certain operational conditions, contribute to a failure.
Failure	the state where a system operates outside its required specification (this can be applied at different levels e.g. system, sub-system, component).
Incident	a significant occurrence or event with potential detrimental consequences.
Accident	an incident with detrimental consequences (due to insufficient control of one or more hazards).

1.2.2   Systems Approach to Risk Control

In practice, plant risks can only be controlled using a systems approach. This provides the overall context in which the computer-based system will operate. The issue of safety of the computer sub-system must be tackled at each step of the life cycle of the plant as a whole, from design to use, eventual modification and dismantling. Figure 1 illustrates the relationship of the target system to the overall system development process.

The process starts with the creation of a plant model which describes the intended functions of the plant within its environment. At this stage, design for safety requires a clear descriptions of what functions the plant must perform under what conditions – the *DO*'s, as well as what the plant must not do under any conditions – the *DON'T*'s. Together the *DO*'s and the *DON'T*'s form the safety-related functional specifications of the plant. In addition, the resources (e.g. time, costs, technology, people) available to develop the plant have to be specified. The design of the plant is an optimization process which balances the functionality of the plant against these resource constraints.

The design of the plant control system is part of this overall development process. The design process for the plant control system starts with an assessment of the optimal distribution of the identified plant control functions (the *DO*'s and the *DON'T*'s) over the three basic *means* of implementation:–

- personnel,
- hardware,
- procedures.

Decisions on the trade-off between these means of implementation are made early in the project, followed by a separate development life-cycle for the three different aspects. The *DO*'s and the *DON'T*'s to be implemented by each approach are derived from the *DO*'s and the *DON'T*'s of the overall system. The three development cycles differ from each other; the hardware development process is well established. The design of procedures and the use of personnel are not so straightforward, and are extremely relevant to safety (witness Chernobyl), but will not be elaborated further here.

The development of the plant control and safety systems hardware will be further sub-divided into hardware-only systems (such as pressure relief valves, simple sensor/actuation circuits etc) and programmable computer-based systems containing software.

Once the separate development cycles for hardware, people and procedures have been completed, the various components of the plant and its control system are available for integration. At this stage a review must be performed to assess the operational readiness of the plant; does the combination of people, hardware and procedures assure that the plant *DO*'s and *DON'T*'s are implemented?

Finally, when the plant is commissioned, the operation of the whole system must be monitored and managed in order to:–

- detect deviations from specified modes of operation,
- correct any design flaws that come to light during operation,
- assure that modifications are correctly implemented,
- analyse any incidents in order to prevent a further occurrence.

Any deviations from the original design will require a reassessment of the operational readiness of the plant (combination of people, hardware and procedures) to assure that the plant hazards are kept under control.

All omissions and oversights in the overall plant life-cycle are management factors which may lead to subsequent accidents. When accidents do occur, this information on the cause should be analysed in detail and the information gained should be used as the basis for design improvements in existing and future plant designs.

1.2.3 Risk Control Strategies

In specifying and developing the plant control system, the following strategies can be used to enhance safety (the approaches are illustrated in Figure 2). Each strategy provides a barrier against the occurrence of a failure in the target system. Even when the computer control system fails, additional hardware protection, manual operations and procedures may be invoked. If these additional protection facilities fail then an 'incident' may occur (e.g. an aircraft engine catches fire).

This may lead to an accident (e.g. if the fire cannot be extinguished).

This document is principally concerned with the implementation of the computer-based components of the overall system. The safety strategies that are applicable to such a target system are discussed below:–

Safety Analysis This is required in the early stages to analyse the safety of the plant and to identify safety-critical plant functions. If probabilistic methods are used, design targets for the reliability and availability of the computer subsystems will be set. At a later stage, safety analysis techniques may be used on the systems implemented, to estimate both the likelihood of computer failure and the impact of different types of failure upon the plant.

Fault Avoidance It is generally accepted that a higher quality and cheaper product can be produced if design faults can be avoided altogether rather than removing them later. There are many methods for minimizing such design faults, ranging from management controls to formal methods.

Fault Detection If faults are introduced into the design, then it is desirable to detect and remove them as quickly as possible. This chapter deals with various methods of fault detection which are applicable at the design phase.

Failure Detection Residual design faults may be present in the final product. In addition random hardware failures will occur during operation. In order to mitigate the effects of these faults it is first necessary to identify that a fault is present. This may be revealed by explicit additional checks or run-time traps which attempt to locate faults within the system by detecting deviations from normal behaviour.

Failure Containment If a failure occurs in the system then the computer system must be designed to minimize its effects. Failures can occur at different levels (a component, sub-system or system). Low-level failures can propagate to affect systems at higher levels unless the lower-level failures are detected and contained. In environments where it is unsafe for the computer system to stop working (such as flight-control computers) the emphasis is placed on fault-tolerance and the continuation of acceptable, possibly degraded, operation. In cases where a safe plant state exists, one can design the computer system to force the plant into this state when failure recovery is infeasible.

Monitoring and Feedback If, despite all control measures, a failure leads to an incident then this should be recorded and analysed in order to review control measures, as well as to learn lessons.

1.3 The Safety Techniques Directory

The design and implementation techniques in this directory are only applicable to the computer-based sub-systems which form part of the overall system development life-cycle. Some of the safety assessment techniques are more general in scope and can be applied at the plant and control system levels.

The majority of the implementation techniques concentrate on the software rather than hardware aspects of target system development (Figure 3). There is also more emphasis on the specification and design of such systems rather than the subsequent verification and validation, although some general material is incorporated to cover these areas.

All the techniques have been grouped into chapters associated with the risk control strategies outlined in the previous section. There is no section on the incident/accident monitoring and feedback as this is considered to be outside the scope of a directory on computer sub-system implementation methods.

Each chapter is prefaced by a list of the techniques to be described. An index is provided at the back of the directory so that specific techniques can be referenced directly. Each technique is described using the following standard format:–

Aim:	A sentence to summarize the main aim of this technique.
Description:	A short description of the means used to meet the stated aims.
Conditions:	Any pre-conditions to be met before the technique can be applied.
Major Advantages:	Reasons for adopting this technique.
Problems or Disadvantages:	Any restrictions on applicability, e.g. problem scale, generality, accuracy, ease of use, cost, availability, maturity etc.
Related Methods:	Alternative, overlapping or complementary techniques.
Assessment:	The EWICS TC7 assessment of this technique. This may include a specific recommendation for its use on a safety-related project.
Tools:	Any tools to support this technique.
References:	References to descriptions of the technique, principally text books and articles in the open literature.

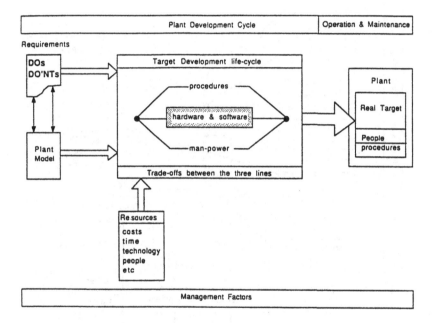

Fig. 1. Target System Life-Cycle

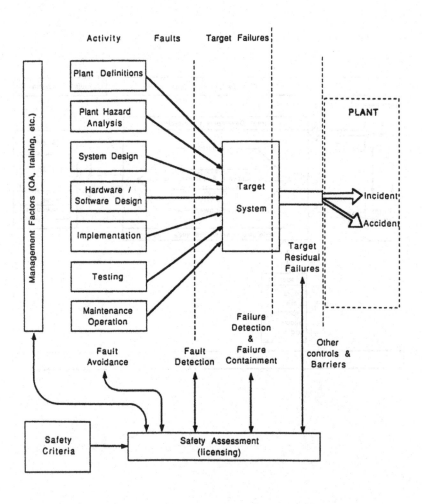

Fig. 2. Strategies for Risk Control

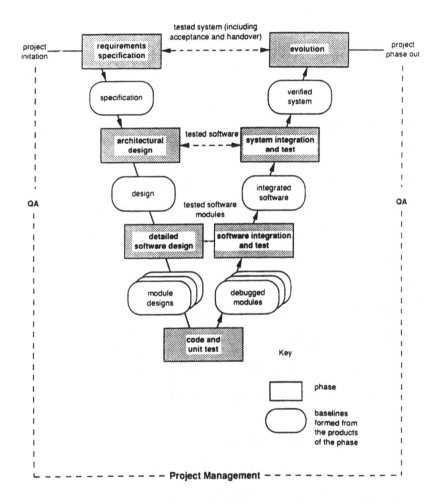

Fig. 3. Software Development Life-cycle

Chapter 2

Safety Analysis

2.1 Cause Consequence Diagrams

2.2 Common Mode Failure Analysis

2.3 Event-Tree Analysis

2.4 FMEA - Failure Modes and Effects Analysis

2.5 FMECA - Failure Mode, Effects and Criticality Analysis

2.6 Fault Tree Analysis

2.7 GO Charts

2.8 HAZOP – System Hazard and Operability Analysis

2.9 Markov Modelling

2.10 MORT - Management Oversight and Risk Tree

2.11 Prediction and Measurement of Software Reliability

    2.11.1 Reliability Growth Models
    2.11.2 Complexity Models

2.12 PHA – Preliminary Hazard Analysis

2.13 PRA – Probabilistic Risk Assessment

2.14 Reliability Block Diagrams

2.15 Sneak Circuit Analysis

2.1 Cause Consequence Diagrams

Aim: To model, in a diagrammatical form, the sequence of events that can develop in a system as a consequence of combinations of basic events.

Description: It can be regarded as a combination of fault-tree and event-tree analysis. Starting from a critical event, a cause-consequence graph is traced backwards and forwards. In the backwards direction it is equivalent to a fault tree with the critical event as the given top event. In the forward direction the possible consequences arising from an event are identified. The graph can contain vertex symbols which describe the conditions for propagation along different branches from the vertex. Time delays can also be included. These conditions can also be described with fault trees. The lines of propagation can be combined with logical symbols, to make the diagram more compact. A set of standard symbols are defined for use in cause consequence diagrams. The diagrams can be used to compute the probability of occurrence of certain critical consequences.

Conditions: No particular conditions.

Major Advantages:

- The method is systematic.
- The method is comprehensive, and is capable of handling most aspects of the development.
- The diagram is easy to understand.
- It is possible to make probability computations on the basis of a cause consequence diagram.
- The method is computer supported.

Problems or Disadvantages:

- The diagram can be very large.

- Difficult to describe continuous development.
- The possibility of rejoining development lines may hide some peculiarities of separate development sequences.

Related Methods: Similar to event trees. Utilizes fault tree analysis.

Assessment: Recommended in assessment of hardware systems. It is more difficult to use in software systems, as some software failures are hard to represent and software failure rates cannot be quantified. However, failures in software modules can be incorporated as basic events in the assessment of hardware/software systems.

Tools: Computer supported construction is possible.

References:

1. *The Cause Consequence Diagram Method as a Basis for Quantitative Accident Analysis.* D.S. Nielsen, Risø-M-1374, 1971.

2.2 Common Mode Failure Analysis

Aim: To identify potential failures in redundant systems or redundant sub-systems which would undermine the benefits of redundancy because of the appearance of the same failures in the redundant parts at the same time.

Description: Computer systems intended to take care of the safety of a plant often use redundancy in hardware and majority voting. This technique is used to avoid random component failures, which would tend to prevent the correct processing of data in a computer system.

However, some failures can be common to more than one component. For example, if a computer system is installed in one single room, shortcomings in the air-conditioning, might reduce the benefits of redundancy. The same is true for other external effects on the computer system such as: fire, flooding, electromagnetic interference, plane crashes, and earthquakes. The computer system may also be affected by incidents related to operation and maintenance. It is essential, therefore, that adequate and well documented procedures are provided for operation and maintenance. Extensive training of operating and maintenance personnel is also essential.

Internal effects are also major contributors to Common-Mode Failures (CMF). They can stem from design errors in common or identical components and their interfaces, as well as ageing components. CMF-Analysis has to search the system for such potential common failures. Methods of CMF-Analysis are general quality control, design reviews, verification and testing by an independent team, and analysis of real incidents with feedback of experience from similar systems. The scope of the analysis, however, goes beyond hardware. Even if 'diverse software' is used in different chains of a redundant computer system, there might be some commonality in the software approaches which could give rise to CMF. Errors in the common specification, for example.

When CMF's don't occur exactly at the same time, precautions can be taken by means of comparison methods between the redundant chains which should lead to detection of a failure, before this failure is common to all chains. CMF analysis should take this technique into account.

Conditions: A thorough understanding of the system to be analysed.

Major Advantages: Covers areas not addressed by other techniques.

Problems or Disadvantages: Problems arise from the fact that CMF analysis cannot be formalized, but always needs human creativity to identify CMF-situations. A resulting disadvantage is the man-hours effort involved in striving for completeness.

Results: Identification of potential CMF's together with quantification where possible (e.g. by using beta factors) and qualification of their effect. From the quantification, the calculated frequencies of common mode failures would be assessable against criteria, if existing, and cost of CMF-effects and counter-measure costs can be balanced. Assessment of defences employed against CMF.

Related Methods: A necessary supplement to the other methods. It profits from a previous Event Tree, or Cause Consequence Analysis, as well as a Failure Modes and Effects Analysis in that a thorough understanding of the system is a prerequisite to CMF Analysis. Identified potential CMF can be further evaluated in these techniques. Probabilistic methods are applicable whenever CMF rates can be determined. In cases where CMF's do not occur simultaneously, probabilistic methods for individual failures can be a substitute to CMF Analysis.

Assessment: The technique is not well developed, but it is necessary to apply CMF Analysis. Without consideration of CMF's, the reliability of redundant systems would be overestimated.

Tools: No tools (besides those of Fault Tree and Event Tree Analysis which can display effects of CMF already conceived in another way).

References:

1. Bibliography by Balbir S. Dhillon, Microelectronic Reliability, Vol. 18, Pergamon Press (1979) 0026-2714 78 0533 502.00/0.

2. *Review of Common Cause Failures.* I.A. Watson, UK-AEA, National Centre for Systems Reliability, Wigshaw Lane, WA3 4NE, England, NCSR R 27, July 1981.

3. *Common-Mode Failures in Redundancy Systems.* I.A. Watson and G.T. Edwards, Nuclear Technology, Vol. 46, Dec. 1979.

4. *A Study of Common-Mode Failures.* G.T. Edwards and I.A. Watson, UKAEA, National Centre for Systems Reliability, Wigshaw Lane, WA3 4NE, England, SRD R 146, July 1979.

5. *Procedures for treating Common Cause Failures in Safety and Reliability Studies.* A. Mosleh et al, NUREG/CR-4780, PLG-0547, Nov 1987.

2.3   Event-tree Analysis

Aim: To model, in a diagrammatical form, the sequence of events that can develop in a system after an initiating event, and thereby describe how serious consequences can occur.

Description: On the top of the diagram one writes the sequence conditions that are relevant in the development following the initiating event which is the target of the analysis. Starting under the initiating event, one draws a line to the first condition in the sequence. There the diagram branches off into a 'yes' and a 'no' branch, describing how the future developments depend on the condition. For each of these branches one continues to the next condition in a similar way. Not all conditions are, however, relevant for all branches. One continues to the end of the sequence, and each branch of the tree constructed this way represents a possible consequence. The event tree can be used to compute the probability of the various consequences based on the probability and number of conditions in the sequence.

Conditions: A knowledge of the potential impact of events on the system.

Major Advantages:

- The tree is easy to draw when the sequence of events is established.
- The tree is easy to understand.
- It is easy to make probability computations on the basis of an event tree.

Problems or Disadvantages:

- It may be difficult to identify the complete sequence of conditions.
- It is difficult to take into account different failure modes.

- It is difficult to take into account dependent failures, common equipment, and common fault interaction.
- The event tree can become very large.

Related Methods: Similar to cause-consequence diagrams. Complementary to fault tree analysis.

Assessment: Recommended in conjunction with fault tree analysis as an alternative to cause-consequence diagrams.

Tools: Some automated tools do exist.

References:

1. *Event Trees and their Treatment on PC Computers.* N. Limnious and J.P. Jeannette, Reliability Engineering, Vol. 18, No. 3, 1987.

2. *PRA Procedures Guide.* NUREG/CR–2300, Division of System Safety Office of Nuclear Reactor Regulation U.S. Nuclear Regulatory Commission Washington, D.C. 20555, 1983.

2.4   FMEA – Failure Modes and Effects Analysis

Aim: To systematically analyse the components of the target system with respect to certain attributes relevant to safety assessment.

Description: The system is divided into units to facilitate analysis. Each unit will consist of a number (one to several hundreds) of components whose failure modes are known or can be predicted. Each failure mode of each component is considered in isolation, and the effect of that failure on the system is analysed. Coincident failures are not considered *unless* one failure can cause a secondary failure. The set of component failure modes which result in a particular system output failure mode can be quantified if the data on the discrete component failures are available.

Standardized forms exist in order to facilitate the analysis.

Conditions: A full system description is needed.

Major Advantages:

- The results constitute an essential input to fault tree analysis and similar numerical methods.
- The method is systematic.
- The method is supported by standardized forms to complete.
- The method permits an analysis of the capability for detecting component failures.

Problems or Disadvantages:

- A comprehensive FMEA may be very time consuming and expensive.

- It is carried out after design, and so is too late to influence design changes.
- It assumes extreme failures.
- The failure modes (and percentage of the failure rate for each mode) are practically impossible to determine for integrated circuits.
- The method is inefficient, since *all* components are analysed, even when their failure does not affect the output.

Related Methods: The results can be utilized in fault tree analysis and event tree analysis.

Assessment: Recommended in all system reliability analyses.

Tools: Standardized diagrams. Computer supported.

References:

1. *System Reliability Engineering Methodology: A Discussion of the State of the Art.* J.B. Fussel and J.S. Arend, Nuclear Safety 20(5), 1979.

2. *Fault Tree Handbook.* W.E. Vesely et. al., NUREG–0942, Division of System Safety Office of Nuclear Reactor Regulation U.S. Nuclear Regulatory Commission Washington, D.C. 20555, 1981.

3. *Reliability Technology.* A.E. Greene and A.J. Bourne, Wiley-Interscience, 1972.

4. *Reactor Safety Study — An Assessment of Accident Risks in US Commercial Nuclear Power Plants.* WASH–1400, NUREG–75/104, Division of System Safety Office of Nuclear Reactor Regulation U.S. Nuclear Regulatory Commission Washington, D.C. 20555, 1975.

2.5   FMECA – Failure Mode, Effects and Criticality Analysis

Aim: To rank the criticality of components which could result in injury, damage or system degradation through single-point failures in order to identify those components which might need special attention and control measures during design or operation.

Description: Criticality can be ranked in many ways. The most laborious method is described by the Society for Automotive Engineers in ARP 926. In this procedure, the criticality number for any component is indicated by the number of failures of a specific type expected during each million operations occurring in a critical mode. The criticality number is a function of nine parameters, most of these have to be measured. A very simple method for criticality determination is to multiply the probability of component failure by the damage that could be generated: this method is similar to simple risk factor assessment.

Conditions: A detailed system description of functions, conditions and (hardware) components is needed.

Major Advantages: Can be used for evaluation of system components of which a failure may result in a potential loss of life, in loss of system functioning or availability, or in excessive unscheduled maintenance.

Problems or disadvantages: The detailed criticality ranking method as described by the SAE requires a great deal of effort. In simplified methods, the essential criticality attribute "potential damage" may be missing. For new, unique systems, component failure data will be less than adequate for a quantitative criticality ranking. A specific component may have several failure modes with only certain ones possibly causing safety problems or harm.

Results:

- Helps to identify components which require more intensive analysis regarding hazard control or fail-safe design.
- Helps to identify components which need tight quality control, protective handling, etc.
- Helps to assess special requirements to be included in the specifications for suppliers.
- Helps to establish measurable quality attributes and acceptance criteria for critical components.
- Helps to identify where hazard control resources could be applied most effectively.

Related Methods: FMECA is an extension of FMEA. Logic Trees, network analysis and (Preliminary) Hazard Analysis can also be used to determine items that would be critical.

Assessment: Recommended for safety critical discrete hardware systems where reliability data of the components is available. The FMECA technique has become less relevant since complex systems are based on linked integrated subsystems, and other risk assessment techniques were developed, e.g. HA-ZOP.

Tools: not known, tools for quantitative fault tree or network analysis may be useful.

References:

1. *Design Analysis Procedure for Failure Modes, Effects and Criticality Analysis (FMECA).* Aerospace Recommended Practice (ARP) 926, Society of Automotive Engineers (SAE), USA, 15 September 1967.

2. *Requirements for Reliability Programs (for Systems and Equipment).* MIL-STD-785, DOD, USA.

3. *Handbook of System and Product Safety.* W. Hammer, Prentice Hall, Englewood Cliffs, N.J., 1972, ISBN 0-13-382226-5.

2.6 Fault Tree Analysis

Aim: To aid in the analysis of events, or combination of events, that will lead to a hazard or serious consequence.

Description: Starting at an event which would be the immediate cause of a hazard or serious consequence (the 'top event') analysis is carried out along a tree path. Combinations of causes are described with logical operators (and, or etc.). Intermediate causes are analysed in the same way, and so on back to basic events where analysis stops.

The method is graphical, and a set of standardized symbols are used to draw the fault tree. It is mainly intended for the analysis of hardware systems, but there has also been attempts to apply this approach to software failure analysis.

Conditions: Cause and effect of events must be known or calculable.

Major Advantages:

- The method is systematic.
- It is well suited as a basis for the computation of the probability of the top event, if one knows the probabilities of the basic events. There are a number of computer application packages which provide this computation.
- There is available computer support to construct fault trees.
- There is wide experience in the use of this method in safety analysis.

Problems or Disadvantages:

- The fault tree may get very large and complex.
- Dynamic aspects and time aspects are not addressed particularly well.

- The method concentrates its attention to specific top events, and is therefore not well suited to reveal other serious consequences.

Related Methods: Similar to the block diagram method. Uses failure modes and effects analysis (FMEA). Used in cause-consequence analysis.

Assessment: Recommended for system safety analysis and HAZOPS.

Tools: Markov modelling tools. Various application tools for construction of fault trees and for reliability assessment based on fault trees are available (e.g. ORCHARD, Safety and Reliability Directorate, Wigshaw Lane, Culcheth, WA3 4NE, England).

References:

1. *System Reliability Engineering Methodology: A Discussion of the State of the Art.* J.B. Fussel and J.S. Arend, Nuclear Safety 20(5), 1979.

2. *Fault Tree Handbook.* W.E. Vesely et. al., NUREG–0942, Division of System Safety Office of Nuclear Reactor Regulation U.S. Nuclear Regulatory Commission Washington, D.C. 20555, 1981.

3. *PRA Procedures Guide.* NUREG/CR–2300, 1983.

4. *Reliability Technology.* A.E. Greene and A.J. Bourne, Wiley-Interscience, 1972.

5. *Automatic Fault Tree Construction with RIKKE – A Compendium of Examples.* J.R. Taylor, Risø-M-2311, Risø National Laboratory, DK–4000, Roskilde, Denmark, 1981.

2.7 GO Charts

Aim: For reliability analysis of complex systems (including components with two or more failure modes) mainly during the design stage.

Description: Seventeen fundamental operators are available to model the system components. Utilizing the library of operators available a GO graph is constructed for each application. Each component of the system to be analysed is modelled by an operator. Each operator allows a particular class of components to be described by the following information:

- input signals to the component.
- operational mode.
- output signals from the component.

Following the structure of the system, the inputs and outputs of different operators are connected, to construct the GO graph which mirrors the original structure. Finally, a probability is associated with the different states of the operators, and the graph is executed for reliability assessment.

Conditions: The system components must be statistically independent.

Major Advantages: Reduction in the time spent for the analysis of complex systems with multiple-failure-conditions. Easy to use, flexible and capable of modelling.

Problems and Disadvantages: The seventeen operators used by the GO methodology may not represent all the failure modes of all components of a real system. In consequence, it can only be used for a preliminary analysis, and different reliability models are necessary to deal with more complex fault models. It is poor from a quantitative viewpoint.

Results: Qualitative analysis of complex systems with multi-state components.

Related Methods: Should be complemented with a fault tree or Markov technique for a quantitative reliability analysis.

Assessment: Recommended for a qualitative analysis during the design stage.

Tools: Some programs for automatic construction and analysis.

References:

1. *An Introduction to GO.* B. Gateley, Kaman Science Corporation, K-75, 94U(R), 1975. Kaman Science Corporation, 1500 Garden of the Gods Road, Colorado Springs, Colorado 80907.

2. *System Reliability Analysis Using the GO Methodology.* R.L. Williams and W.Y. Gateley, Kaman Science Corporation K-77-15U(R), 1977.

3. *Use of GO Methodology to Determine Fault Sets.* R.L. Williams and W.Y. Gateley, Kaman Science Corporation K-77-21U(R), 1977.

4. *Computerized Probability Risk Assessment.* T. Jackson, Proc./ Fifth European Symposium on System Safety, Cannes, April 1988.

2.8 HAZOP – System Hazard and Operability Analysis

Aim: To establish by a series of systematic examinations of the 'component' sections of the computer system and its operation, failure modes which lead to potentially hazardous situations in the controlled system (the plant).

Typical hazardous events on the controlled systems are fire, explosion, release of toxic material (chemical or nuclear) or serious financial loss.

It is assumed that the hazardous events on the system being controlled have been identified in a separate Hazard Analysis and the consequence of the hazardous events classified into degrees of seriousness.

The analysis which covers all stages of the project life-cycle (from specification through design to maintenance and modification) is intended to identify at each stage, events or failure modes which could lead to potential hazards and thus eliminate them.

Description: The analysis is carried out by a team of engineers (covering computer, instrument, electrical, process, safety and operational disciplines) led by a trained specialist in hazard analysis techniques in a series of scheduled meetings.

It is important that a fixed time schedule is allocated within the project for the meetings — each one scheduled for at least half a day and for effectiveness no more than four per week to allow for maintaining the flow of accompanying documentation.

Prior to the study, agreed checklists for the systematic examination will have been compiled and for each section of the system leading questions are asked namely: what if it happens? how can it happen? when can it happen? does it matter? From positive answers further questions are asked, for example: what can be done about it? when must it be done? is there an alternative? etc.

For the first part of the analysis it is suggested that the computer installation as a whole is examined. The second and subsequent parts involve detailed examination of the relevant parts of the computer systems itself.

At each part or stage of the analysis, the objective is to identify failure modes which lead to potential hazards on the controlled system and the degree of their effect. The major component parts of the computer system listed above are further sub-divided and where necessary subjected to a separate analysis.

At suitable points throughout the systematic analysis, action review meetings will be arranged.

It is essential that comprehensive records of the meetings are kept, for they will form a substantial part of the system hazard/safety dossier.

After a number of meetings it is suggested that a review meeting be held to ensure that actions are followed up, modifications suggested during the Study meetings are incorporated into the study etc.

Conditions:

- This technique needs a trained specialist in hazard analysis techniques to lead the team of engineers who have the knowledge of the 'system' being controlled and the computer system being used.

- Time is allocated to carry out the meetings on a regular basis.

- Formal reviews are carried out throughout the project.

Major Advantages:

- It identifies potential hazards at every stage of the project life-cycle and their effects.
- It allows the effect of failure to be eliminated or reduced to an acceptable level at every stage of the project.
- It highlights the need for any special procedures or corrections to existing ones.
- It may help the design team to decide on the degree of redundancy and diversity required.
- It may contribute to the software specification, highlighting the need for special routines, modifications, or deletion of routines etc.
- It focuses on the sensitive areas of the system which, on failure could lead to potentially hazardous consequences and this highlights where effort can be most effectively applied to improve the system safety, reliability and availability.
- It provides a series of documents which will form part of the system safety dossier.

Problems or Disadvantages:

- It takes time to carry out the study and with the expertise required will increase the cost of the project.
- It needs a rigid scheduling of meetings to be cost effective — possibly at times within the project when this is not easy to achieve.
- It needs someone of sufficient technical ability to maintain the documentation flow.

Results:

- Identification of failure modes.
- Classification of effect of potential hazards.
- Concise documentation.
- Agreed actions and modifications.

• Information for further analysis, i.e. quantification.

Related Methods: Will provide information for input to FMEA, FTA etc.

Assessment: This technique should be considered mandatory for safety related systems.

Tools: No specific tools.

References:

1. *HAZOP and HAZAN*, T.A. Kletz, 1986, 2nd Edition, Institution of Chemical Engineers, 165-171 Railway Terrace, Rugby, CV1 3HQ, UK.

2. *Reliability and Hazard Criteria for Programmable Electronic Systems in the Chemical Industry*. E. Johnson, in Proc. of 'Safety and Reliability of PES', PES 3 Safety Symposium, B.K. Daniels (ed.), 28–30 May 1986, Guernsey, Channel Islands, Elsevier Applied Science, 1986.

3. *Reliability Engineering and Risk Assessment*. Ernest J. Henlty and Hiromitsu Kumamoto. Prentice Hall 1981.

4. *Systems Reliability and Risk Analysis*. Ernst G. Frenkel. Martinus Nijhoff, 1984.

5. *Software Safety*. N.G. Leveson, in 'Resilient Computer Systems' T. Anderson (Ed.), Collins, 1985.

6. *Software Safety in Computer Controlled Systems* N.G. Leveson, Computer, Feb. 1984.

7. *Risk Assessment of Computer Controlled Systems*. M.O. Fryer, IEEE Transactions on Software Engineering SE-11, Jan. 1985.

2.9 Markov Modelling

Aim: To evaluate the reliability or safety or availability (i.e. dependability attribute) of a system.

Description: A graph of the system is constructed. The graph represents the status of the system with regard to its failure states (the failure states are represented by the nodes of the graph). The edges between nodes, which represent the failure events or repair events, are weighted with the corresponding failure rates or repair rates. Note that the failure events, states and rates can be detailed in such a way that a precise description of the system is obtained, e.g. detected or undetected failures, manifestation of a larger failure, etc.

The Markov technique is suitable for modelling redundant systems in which the level of redundancy varies with time due to component failure and repair. Other classical methods, e.g. FMEA and FTA, cannot readily be adapted to modelling the effects of failures throughout the life-cycle of the system since no simple combinatorial formulae exist for calculating the corresponding probabilities.

In the simplest case, the formulae which describe the probabilities of the system are readily available in the literature or can be calculated manually. In more complex cases, some methods of simplification (absorbing states...) exist and at least very complex cases can be calculated by computer simulation of the graph.

Conditions: Knowledge of the rates (e.g. failure rate, repair rate, etc.); at least conservative estimates. For more complex cases, an appropriate simulation program which would have to be obtained.

Major Advantages: Markov Modelling provides an easy way to estimate the dependability of systems in which the level of redundancy varies during the life-cycle of the system.

Parametric studies can be performed to assess the impact of:

- different system configurations,
- different rates,
- different repair strategies.

Problems or Disadvantages:

- Hard to solve analytically, a Markov simulation is necessary for most practical applications.
- Depending on the simulation method of the program, some results may not be truly representative when the rates differ by orders of magnitude (for example, failure rate $= 10^{-6}h^{-1}$ and repair rate $= 1h^{-1}$).
- The number of states may become very large.

Related Methods: The method is probabilistic. It can be used as a complement to other analysis methods: Failure Modes and Effects Analysis, Fault Tree Analysis, Cause-Consequence Diagrams.

Assessment: Recommended for dependability evaluation of redundant hardware. A standard method in these cases.

Tools: Tools exist to simulate or compute Markov graphs.

References:

1. *The Theory of Stochastic Processes*. R.E. Cox and H.D. Miller. Methuen and Co. Ltd. London, UK, 1968.

2. *Finite MARKOV chains*. J.G.Kemeny and J.L. Snell. D. Van Nostrand Company Inc. Princeton, 1959.

3. *Reliability Handbook* B.A. Koslov and I.A. Ushakov. Holt Rinehart and Winston Inc. New York, 1970.

4. *The Theory and Practice of Reliable System Design*. D.P. Siewiorek and R.S. Swarz. Digital Press. 1982.

2.10   MORT – Management Oversight and Risk Tree

Aim: To manage risks in an organisation, using a systemic approach, in order to increase reliability, assess risks, control losses and allocate resources effectively.

Description: MORT is an acronym that carries two primary meanings: (a) the MORT "tree" or logic diagram, which organises risk, loss, and safety programme elements and is used as a master work sheet for accident investigations and programme evaluations.; and (b) the total safety programme as a subsystem to the major management of an organisation. The philosophy underlying MORT can be summarised as follows:

- management takes risks of many kinds;
- risks in one area effect operations in other areas;
- risks should be made explicit where practicable;
- risk management tools should be flexible enough to suit a variety of diverse situations.

The MORT process includes four main analytical tools. In Change Analysis, a problem-free situation is compared with a problem situation in order to isolate causes and effects of change; this tool is based upon the Kepner-Tregoe method of rational decision making.

Energy Trace and Barrier Analysis is based on the idea that uncontrolled energy flows in the absence of adequate barriers can cause accidents; it encompasses the simple energy-barrier-target concept which underlies the question "What happened" in case of an accident investigation.

The Positive (Success) Tree Design reverses the logic of fault tree analysis and shows all that must be performed and the proper sequencing of events needed to accomplish an objective.

The MORT Tree Analysis combines principles from the fields of management and system safety, using fault tree methodology. It helps to discover *what* happened and *why* in case of an accident. The MORT tree organises over 1500 basic events (causes) leading to 98 generic events (problems). Both specific control factors and management system factors are analysed for their contributions to a specific accident. People, procedures and hardware are considered separately, then together, as key system safety elements. A MORT tree analysis is performed by a qualitative evaluation of underlying events (given by the tree) of potential problems. A potential problem becomes an actual problem if any of the underlying events is to be judged as "less than adequate" in relation to the stated top event. The route towards such an event is the key to a practical solution to the main problem: prevention of actual losses or assurance of a system's objective. The MORT tree consists of two major branches. In the S-branch a systemic reconstruction of the physical conditions of an accident is produced. The M-branch covers all the basic organisational functions, including all design cycle functions, which set the constraints for a functional operation. Therefore, the S-branch includes transfer points towards lower-tier branches and sub-trees in the M-branch. If used for programme evaluation, only the M-branch is applicable.

Conditions: MORT requires simultaneous consideration and development of people, procedures and hardware and clearly defined system goals (programme evaluation) or clearly stated losses (accident analysis). It also requires a relatively closed (organisational) system in which the desired operations have to take place.

Major Advantages: The MORT tree is the only single, comprehensive accident analysis tool which relates circumstantial causal factors with organisational control factors. It can be used in all points of a system's life cycle.

Problems or disadvantages: The MORT tree has been applied to complex, highly technological production systems, e.g. nuclear plants. The tree is too comprehensive for relatively simple production systems. A simplified MORT tree requires revalidation before it can be applied successfully.

Results: It enables an experienced MORT tree analyst to pinpoint all relevant causal accident factors and produce valid options for assurance of accident prevention within a significantly shorter time period than possible through other analysis methods. The Department of Energy (DOE, USA) requires MORT training and use by its contractors for accident/ incident investigations and uses MORT to assess its contractor safety programmes.

Related Methods: although the MORT tree is constructed like a fault tree, there is an essential difference: the MORT tree is a generic events tree that does not need to be constructed for each specific purpose and which enables the use of lower tier sub-trees for partial programme review. The other methods are supporting tools in a MORT-tree analysis which can be applied separately.

Assessment: Useful in project planning, functional specification of a target (sub)system, accident/incident analysis and safety programme evaluation.

Tools: Computerised MORT tree (CMORT), Computerised Operational Readiness Tree (CORT), C-CAUSE (computerised events and causal factors analysis tool), CALT/PALT (Computer Aided Logic Tree for qualitative and quantitative fault tree analysis)

References:

1. *MORT Safety Assurance Systems*. William G. Johnson. Marcel Dekker Inc., New York, Basel. 1980. (ISBN 0-8247-6897-3).

2. *MORT Users Manual.* D.S. Conger, K.J. Elsea. EG and G Services, Inc., Woodstock, GA 30188, USA. 1987.

3. *Manuals.* SSDC-1 to SSDC-25, detailing how to perform quick or in-depth analysis of different parts of an organisation, e.g. maintenance department. System Safety Development Center, EG and G Services, Inc. Idaho Falls, Idaho 83415, USA. 1975-1982.

2.11 Prediction and Measurement of Software Reliability

This section reviews the techniques for the predictions and measurements of software reliability in high integrity industrial applications.

On the opposite side of hardware components theory, where well-tried techniques are now agreed by most users, no universal method has been accepted to quantify the confidence one can have in a software product.

Many models have already been proposed: all of them suggest a set of mathematical equations to estimate program performance such as reliability, number of remaining errors or Mean Time To Failure.

All these models *cannot be considered as certification tools*; on the other hand, they may be very useful for rough estimation and project management.

These methods can be classified in three classes:

Reliability Growth Models – which tend to estimate the reliability of software during the testing phase by providing a measure of failure rates;

Complexity Models – which address different aspects of static complexity of programs;

Help to software project management – these are models with very specific goals, such as estimation of development cost. This third class is very far from an assessment technique and is not described here.

2.11.1   Reliability Growth Models

Aim: To predict the current software failure rate and hence the operational reliability.

Description: Two kinds of models are to be distinguished here:

- models which tend to estimate the number of remaining errors in a software product, and hence minimum time to correct these bugs. These models are known as Musa models, Jelinsky–Moranda models or Bug-counting models;

- models which tend to provide the current failure rate of a program, and hence minimum time required to reach a certain reliability; most famous examples are the Littlewood, and the Littlewood–Verrall models.

In the first case, the failure rate of the software is assessed to be a function of the number of remaining errors (e.g. a sum of gamma functions for Littlewood model) : the aim is to estimate the corresponding parameters.

Conditions: Development history of the particular software is required.

Major Advantages: Means of obtaining a quantitative opinion about the results of the software development phase.

Problems or Disadvantages: The diversity of reliability models prevents users from having contractual confidence in the results. Moreover, the usual assumptions made in order to run these models (independence of faults, randomness of test samples, no adjunction of bugs when correcting, etc) are too far from the real software development conditions.

Related Methods: Related to testing methods.

Assessment: There are currently many models with few convincing validations. There are severe problems in using them to predict the levels of reliability (and safety) which high integrity application areas demand. Nevertheless, they do have a role:

- for forming a general opinion about the reliability level of the software product;

- as tools to compare different software modules under the same project conditions;

- as an aid for project management, e.g. deciding when to change testing phases.

Tools: Program packages are available for both Musa and Littlewood type models (Prof. B. Littlewood, City University, Northampton Square, London EC1V OHV).

References:

1. *Evaluation of competing software reliability predictions.* A.G. Abdalla, P.Y. Chan, and B. Littlewood. IEEE Transactions on Software Engineering, Vol. SE-12, No. 9, Sept. 1986.

2. *Software reliability status and perspectives.* C.V. Ramamoorthy and F. Bastani. IEEE Transactions on Software Engineering, Vol SE-8, 1982.

3. *Software reliability models: a review.* J.G. Shantikumar. Microelectr. Reliability, Vol. 23, No. 5. 1983.

4. *Dependability evaluation of software systems in operation.* J.C. Laprie. IEEE Transactions on Software Engineering, Vol. SE-10, No. 6, Nov. 1984.

5. *Dependability prediction: comparison of tools and techniques.* M. Mulazzani, K. Trivedi, IFAC SAFECOMP 86, W. Quirk (ed.), Pergamon Press, Oxford, 1986.

6. *Software dependability of a telephone switching system.*
   K. Kanoun and T. Sabourin. Proceedings of Fault-
   Tolerant Computing Symposium (FTCS-17), IEEE, July
   1987.

7. *Techniques for the Verification and Validation of Safety
   Related Software.* in 'Dependability of Critical Computer Systems: Part 1', Elsevier Applied Science, 1988.

2.11.2 Complexity Models

Aim: To predict the reliability of programs from properties of the software itself rather than from its development or test history.

Description: These models evaluate some structural properties of the software and relate this to a desired attribute such as reliability or complexity. The assumption is that some property, such as measured complexity, is obviously correlated to reliability, i.e. more complex programs are more likely to be unreliable. Software tools are needed to evaluate most of the measures. We give here a summary of these methods:

- Graph Theoretic Complexity: this measure can be applied early in the life-cycle to assess trade-offs, and is based on the complexity of the program control flow graph, represented by its cyclomatic number.

- Number of ways to activate a certain module (accessibility): the more a module can be accessed, the more likely it is to be debugged.

- Software Science: this measure computes the 'program length' by counting operators and operands. It provides a measure of complexity and estimates development resources. Based on Halstead's work.

- Number of entries and exits per module: minimizing the number of entry/exit points is a key feature of structured design and programming techniques.

Conditions: Tool support is necessary. Most measures are directed towards a computer language, but some also work on specifications.

Major Advantages: The use of structural measures allows a direct evaluation of a particular attribute. It enables early measures on a program (as soon as conception) whereas reliability models need time-dependent data.

Problems or Disadvantages: There is no objective correlation of the method results with reliability data. The observed correlations of complexity with detected faults tend to be project- and site-specific.

Related Methods: Related to structured programming.

Assessment: Complexity measures can be used at the design, coding and testing phase to improve the quality of the delivered software by the early identification of over-complex modules and by indicating the level of testing required for different modules. The acceptable levels of complexity cannot be stated in absolute terms, but within particular projects and organizations valid local correlations can be obtained. These measures provide evidence of software quality and hence confidence in the reliability of the product. There are many theoretical and practical problems in relating them to quantified reliabilities.

Tools:

- Logiscope^{TM} (Verilog, rue Nicolas Vauquelin 31081 Toulouse, Cedex, France).
- Spade^{TM} (Program Validation Ltd., 34 Basset Crescent East, Southampton SO2 3FL, England).

References:

1. *A complexity measure*, T.J. McCabe, IEEE Transactions on Software Engineering. Vol. SE-2, No. 4. Dec. 1976.

2. *Models and measurements for quality assessment of software.* S.N. Mohanty. A.C.M. Computing Surveys, Vol. 11, No. 3, Sep. 1979.

3. *Elements of software science.* M. Halstead. Elsevier / North Holland, New York, 1977.

2.12   PHA – Preliminary Hazard Analysis

Aim: To determine during system concept or early development the hazards that could be present in the operational system in order to establish courses of action.

Description: PHA is a cursory review of safety problems related to the system (or plant) goals and to its basic performance requirements, including the environments in which the operations will take place. For each identified primary hazard that could generate injury, damage or loss of system function, the following attributes are to be assessed: cause, effect, hazard category, and potential corrective or preventive measures.

The simplest type of PHA consists of a form in which primary hazards are listed along with their respective causes, effects, categories and corrective measures. In order to ensure that all possible hazards are listed and evaluated, logic tree models of the plant or system have to be developed, in which the top event is stated as a safety problem.

Hazard categories are to be defined in advance, e.g. according to MIL STD-882A:

category I: negligible   will not result in personnel injury or system damage

category II: marginal   can be counteracted or controlled without injury to personnel or major system damage

category III: critical   will cause personnel injury or major system damage, or will require immediate corrective action for personnel or system survival

category IV: catastrophic   will cause death or severe injuries to personnel, or system loss.

The merits of this kind of categorisation are doubtful if much time is wasted to establish that a particular problem is in a low hazard category when it would be better to try to eliminate or minimise the problem.

In most cases, the following basic steps are undertaken for a PHA:

- review of problem known through past experience on similar systems to determine whether they could also be present in the system under development;
- review system goals and basic performance requirements, including the environments in which operations will take place;
- determine the contributory and initiating hazards that could cause or contribute to the primary hazards listed;
- review possible means of eliminating or controlling the hazards according to an established hazard control strategy;
- analyse the best method of amelioration, including damage restriction, in case there is a loss of control of a hazard;
- indicate who is to decide on corrective measures and who is to take the actions.

Conditions: PHA requires information on intended system operation: goals, functions, performance capabilities, operational sequences, environments of operation, and relevant codes, regulations, standards, etc. PHA is of special importance where there is little similarity to previous systems so that experience on hazards is lacking. It should be followed by detailed analysis later in the design cycle.

Major Advantages: Usually, there is little information on design details and even less on procedures available at the stage in which PHA is to be performed: consequently, PHA focusses on the potential major problems at the future, operational system/plant level and can be performed in a relatively short period of time.

Problems or disadvantages: PHA is less useful for systems which are similar to those already in operation and from which much experience has been gained. PHA was first required as a review of hazards in missile systems, i.e. high-energy systems which contain a lot of highly hazardous features. There is not much experience in lower-energy applications.

Results:

- It helps to develop guidelines and criteria to be followed in the design and development process of a system;
- It indicates actions for hazards elimination or control from the beginning of the system life cycle;
- It can be used to designate management and technical responsibilities for safety tasks during the whole system life cycle;
- It can indicate the information that must be reviewed in documents, e.g. standards and specification guidelines, governing precautions and safeguards to be taken for each hazard.

Related Methods: Will provide information for input to detailed analysis (FMEA, FTA, etc.) and project management (decision on hazard control strategies, allocation of responsibilities). If applied in the earliest stage of the system life cycle, the Hazard Operability Analysis techniques can be considered for use instead of PHA.

Assessment: Should be considered for specification of systems which are not similar to those already in operation and from which much experience has been gained.

Tools: Tools for Fault Tree Analysis may be useful.

References:

1. *Handbook of System and Product Safety.* W. Hammer, Prentice Hall, Englewood Cliffs, N.J., 1972, ISBN 0-13-382226-5.

2. *Loss Prevention in the Process Industries.* F.P. Lees, Butterworth, London, 1980, ISBN 0-408-10604.

3. System Safety Program Requirements. Doc.no. MIL-STD-882A, DOD.

2.13 PRA – Probabilistic Risk Assessment

Aim: Evaluates the risks involved in the operation of a safety critical system. The risk assessment forms the basis of design decisions.

Description: Several types of risks are calculated as hypothetical consequences of all possible accidents, namely:

- The number of losses of human lives,
- The number of days reduction in life-span,
- The number of days of illnesses,
- The financial loss,
- The environmental pollution (for example the area damaged) that is additional to the damage related health or financial aspects.

All these aspects constitute the *risk vector*. The calculation considers the contribution to overall risk by each type of hazard. The computation takes into account:

- frequency of a loss-causing event,
- loss in the case of failure,
- frequency of sub-events,
- probability of automated safety system failure per event,
- probability of failure of subsequent manual safety procedures.

On the basis of the calculated risk vector, the design decisions are agreed or revised.

Conditions:

- Detailed knowledge of environment, safety related system, operation, and interaction with plant.
- Enough effort to carry the investigations.

Major Advantages:

- Enables design decisions to be made on a quantitative basis.
- Permits the comparison of possible consequences of particular decisions, with a certain level of confidence.

Problems or Disadvantages:

- Needs considerable effort if the results are to be credible.
- Inexactly known parameters lead to large variances of the results.

Related Methods: Utilises Failure Modes and Effects Analysis, Fault-Tree Analysis, Cause-Consequence Diagrams.

Assessment: Recommended before major design decisions. Not reasonable for the minor system aspects.

Tools: Tools for Fault Tree Analysis may be useful.

References:

1. *Reactor Safety Study — An Assessment of Accident Risks in US Commercial Nuclear Power Plants.* WASH–1400, NUREG–75/104, Division of System Safety Office of Nuclear Reactor Regulation U.S. Nuclear Regulatory Commission Washington, D.C. 20555, 1975.

2. *PRA Procedures Guide.* NUREG/CR–2300, 1983.

3. *Reliability Data Collection and Use in Risk and Availability Assessment.* Editor: H.J. Wingender, Proceedings of the fifth EuReDatA Conference, Heidelberg, April 9–11, 1986, Springer-Verlag.

4. *Risk Analysis in the Process Industries.* Report of the Study Group on Risk Analysis (ISGRA), Intitute of Chemical Engineers, Rugby, UK, 1985.

2.14   Reliability Block Diagrams

Aim: To model, in a diagrammatical form, the set of events that must take place and conditions which must be fulfilled for a successful operation of a system or a task.

Description: The target of the analysis is represented as a success path consisting of blocks, lines and logical junctions. A success path starts from one side of the diagram and continues via the blocks and junctions to the other side of the diagram. A block represents a condition or an event, and the path can pass it if the condition is true or the event has taken place. If the path comes to a junction it continues if the logic of the junction is fulfilled. If it reaches a vertex, it may continue along all outgoing lines. If there exists at least one success path through the diagram the target of the analysis is operating correctly.

Conditions: Conditions and events must be analysable to allow structure of diagram to be implemented.

Major Advantages:

- The method is systematic.
- The diagram is more compact than a fault tree.
- A block diagram may follow the physical structure of the target system quite closely.
- It is suited as a basis for the computation of the probability of the top event, if one knows the probabilities of the blocks. Computer codes are available which make this computation.
- Is is easy to draw a fault tree on the basis of a block diagram.

Problems or Disadvantages:

- It addresses success rather than failures, the latter being more relevant concerning safety.

- It is difficult to take into account different failure modes. This may lead to omission of important conditions.
- The method is not as widely used or as widely computer-supported as fault tree analysis.

Related Methods: Similar to fault tree analysis.

Assessment: Useful for the analysis of systems with relatively straightforward logic, but inferior to fault tree analysis for more complex systems.

Tools: Various tools for reliability assessment based on fault trees.

References:

1. *System Reliability Engineering Methodology: A Discussion of the State of the Art.* J.B. Fussel and J.S. Arend, Nuclear Safety 20(5), 1979.

2. *Fault Tree Handbook.* W.E. Vesely et. al., NUREG-0942, Division of System Safety Office of Nuclear Reactor Regulation U.S. Nuclear Regulatory Commission Washington, D.C. 20555, 1981.

3. *Reliability Technology.* A.E. Greene and A.J. Bourne, Wiley-Interscience, 1972.

2.15   Sneak Circuit Analysis

Aim: To detect an unexpected path or logic flow within a system which, under certain conditions, can initiate an undesired function or inhibit a desired function.

Description: A sneak circuit path may consist of hardware, software, operator actions, or combinations of these elements. Sneak circuits are not the result of hardware failure but are latent conditions inadvertently designed into the system or coded into the software programs, which can cause it to malfunction under certain conditions.

Categories of sneak circuits are:

Sneak Paths which cause current, energy, or logical sequence to flow along an unexpected path or in an unintended direction.

Sneak Timing in which events occur in an unexpected or conflicting sequence.

Sneak Indications which cause an ambiguous or false display of system operating conditions, and thus may result in an undesired action by the operator.

Sneak Labels which incorrectly or imprecisely label system functions, e.g. system inputs, controls, displays, buses, etc., and thus may mislead an operator into applying an incorrect stimulus to the system.

Sneak circuit analysis, relies on the recognition of basic topological patterns with the hardware or software structure (e.g. six basic patterns are identified for software). Analysis takes place with the aid of a checklist of questions about the use and relationships between the basic topological components.

Conditions: The basic design must be available, and broken down into its topological components, an adequate set of questions should be available.

Major Advantages: Applicable to hardware, software, and the combined system. Can be applied to programs written in any language. Directed towards system design flaws rather than component failures.

Problems or Disadvantages: It is labour intensive. It is also likely to be performed late in the development cycle, so changes will be expensive.

Related Methods: Complementary to FMECA and FTA.

Assessment: Should be considered for those components which are safety critical.

Tools: Experimental tools may be available from COSMIC, University of Georgia, 112 Barrow Hall, Athens, Georgia 30602.

References:

1. *Sneak Analysis and Software Sneak Analysis.* S.G. Godoy and G.J. Engels, J. Aircraft, Vol. 15, No. 8, 1978.

2. *Sneak Circuit Analysis.* J.P. Rankin, Nuclear Safety, Vol. 14, No. 5, 1973.

3. *Sneak Circuit Analysis Handbook.* Boeing Aerospace Company, Space Systems Division, Houston, Texas, USA, Doc.no. D2-118341-1, July, 1970

4. *Sneak Circuit Analysis of the AWACS Electrical Power System.* Boeing Aerospace Co. Doc. No. D2-118547-1, Oct 1974.

5. *Application of Software Sneak Analysis to the Terminal Configured Vehicle System.* Boeing Aerospace Co. Doc. No. D2-118594-1, Aug 1976.

Chapter 3

Fault Avoidance

3.1 Avoidance of Complexity

3.2 Certificated Hardware Components

3.3 Certificated Software Components

3.4 Certificated Tools

3.5 Configuration Management

3.6 Development Standards

3.7 Electromagnetic Protection

3.8 Formal Development Methods

    3.8.1 CCS – Calculus of Communicating Systems

    3.8.2 CSP – Communicating Sequential Processes

    3.8.3 Finite State Machines

    3.8.4 OBJ

    3.8.5 Petri Nets

    3.8.6 Temporal Logic

    3.8.7 VDM – Vienna Development Method

    3.8.8 Z

3.9 Formally Designed Hardware

3.10 Information Hiding

3.11 Quality Assurance

3.12 Safe Language Subsets

3.13 Stress Reduction

3.14 Strongly Typed Programming Languages

3.15 Structured Programming

3.16 Structuring the System according to Criticality

3.17 Systematic Analysis and Design Methods

> 3.17.1 CORE – Controlled Requirements Expression
> 3.17.2 JSD – Jackson System Development
> 3.17.3 MASCOT
> 3.17.4 Object-oriented Design and Programming
> 3.17.5 Real-Time Yourdon
> 3.17.6 SADT – Structured Analysis and Design Technique
> 3.17.7 SDL – Specification and Description Language
> 3.17.8 SOM – Systems Development by an Object-oriented Methodology
> 3.17.9 Synchronous Data Flow Specification Languages

3.1   Avoidance of Complexity

Aim: To minimize the chance of error by making the system as simple as possible.

Description: The watchword is KISS "Keep It Simple, Stupid". The basic idea is to critically examine all aspects of the system to check whether it is unnecessarily complex. This process starts at the user requirements level, where all unnecessary functions should be ruthlessly eliminated. It continues at the system design level where the main goal is to minimize effort and originality within the project while maintaining quality. This can be achieved in a number of ways:–

- Looking for the simplest possible alternative for implementing the system *including the possibility of not using a computer at all.*
- Re-use of existing software components and standard systems.
- Isolation of safety-critical functions, so that it requires less effort to implement, verify and certify/license the complete system.

This can be continued into the implementation phase by using a very small and well-knit team to minimise problems of human communication. This will also limit the available resources which helps to ensure that the most straightforward means of implementation are used.

Conditions: A very determined and hard-nosed project manager.

Major Advantages: The final system should arrive earlier and will probably be less expensive. It is likely to be more readily understood and hence be more easily maintained and verified or licensed.

Problems or Disadvantages: The system may have less functionality than the customer ideally wanted.

Related Methods: Inspection methods and static analysis can be used to review the requirements and system design to check for unnecessary complexity.

Assessment: Less frequently used than it should be. Highly recommended for safety critical systems.

Tools: None known.

References:

The Principles of Software Engineering Management. T.Gilb, Addison Wesley, 1987.

3.2   Certificated Hardware Components

Aim: To assure that all hardware components that are used will not reveal inherent weaknesses after their use within the system by screening and segregating the positively certified components.

Description: Virtually all manufactured devices have a life characteristic that is represented by the bathtub curve. This method deals with the first part called the 'infant mortality' period. Specially devised tests are developed and performed to weed out the weak, unreliable components.

They are generally called *Quality tests* and *Screening tests* and *Burn-in*. Quality tests reduce defective items from production lines. Screening tests remove inferior devices, reducing hazard rate by means of stress application. Burn-in tests take the component through the early failure period prior to operational use. The main purpose of these tests is to compress the early failure period and reduce the failure rate to acceptable values as soon as possible.

Conditions: Knowledge of component failure modes, reliability or quality assurance levels, and screening techniques.

Major Advantages: If applied carefully (i.e. in identified high risk areas), certified hardware components can be a cost effective method of obtaining higher reliability. Permits a quantitative assessment of how much 'better' a certain hardware component is with respect to another.

Problems or Disadvantages: It can be very cumbersome and expensive if the system has a high part count, or if very high standards are needed.

Related Methods: It is very tightly coupled with an overall quality assurance program. Also related to configuration management through the guided use of more reliable components if a certain function is particularly 'dangerous'.

Assessment: Recommended for safety-critical systems. In some fields (e.g. Military, Space, Avionics) they are mandatory.

Tools: Automated tools for the computation of system reliability can be used, based on the reliability figures derived for the components.

References:

1. *Electronic Design Handbook.* MIL-HDBK-338.

2. *Test Methods for Electronic and Electrical Component Parts.* MIL-STD-202.

3. *Test Methods and Procedures for Microelectronics.* MIL-STD-883.

4. *Semiconductor Devices General Specifications.* MIL-S-19500.

5. *Failure Rate Sampling Plans and Procedures.* MIL-STD-690.

6. *Microcircuits General Specifications.* MIL-M-38510.

7. *Reliability Assurance Program for Electronic Parts Specification.* MIL-STD-790.

8. *Reliability Prediction of Electronic Equipment.* MIL-HDBK-217.

9. *Parts Control Program.* MIL-STD-965.

3.3   Certificated Software Components

Aim: To minimize the development of new software through the use of existing components of known level of confidence or quality.

Description: A certificated software component is one that has been subjected to some standard degree of verification and validation. Ideally, this will have been carried out against an adequate specification from, say, a national or international standards body or accredited organization.

International certification is carried out, for example, on software supporting the Open Standards Interface (OSI) and the Graphical Kernel System (GKS). Component libraries are commercially available such as the Numerical Algorithms Group (NAG) Library in the UK, but these components are not always libraries of *certificated* components, i.e. components that have been independently validated.

Conditions: The component must be available, its function must be appropriate, well-documented, easy to use and the certification should be of a sufficiently high standard for the purpose.

Major Advantages: The use of a certificated component should give more confidence in its correctness than the development of a new one from scratch since such a component will have been tested to a certain level. Standardization of components should lead to greater familiarity and hence reduce the risk of using incorrect combinations of components. There are subsidiary advantages in that the purchase of a certificated component should be cheaper and quicker than bespoke development.

Problems or Disadvantages: There are no major disadvantages.

Results: Increased confidence that the software will work.

Related Methods: Related to object-oriented design.

Assessment: Wherever possible certificated components should be used, but there are few examples to date. Depending on the quality of the certification process and the levels of confidence they imply, additional validation and verification may be necessary.

Tools:

NAG Library (numerical algorithms). NAG Ltd, Wilkinson House, Jordan Hill Road, OX2 8DR, UK.

The Booch Components (data structures and data manipulation/comparison tools), Wizard Software, 835, S. Moore St., Lakewood, CO 80226, USA.

SLATEC 3.0. (mathematical subroutines), Package Id 0820. National Energy Software Centre, Argonne National Lab., 9700 South Cass Av., Argonne, Illinios, 60439, USA / NEA Data Bank, F–91191 Gif sur Yvette, Cedex, France.

References:

1. *NATLAS (National Testing Laboratory Accreditation Scheme) Software Unit Test Standard and Method.* NAMAS Executive, National Physical Laboratory, Teddington, Middlesex, TW11 OLW, England.

2. *Software Components with ADA - Structures, Tools and Subsystems.* G. Booch, The Benjamin/Cummins Publ. Corp. Inc., Menlo Park, California, 1987.

3.4   Certificated Tools

Aim: Tools are necessary to help developers in the different phases of software development. Wherever possible tools should be certificated so that some level of confidence can be assumed regarding the correctness of their outputs.

Description: A certificated tool is one that has been determined to be of a particular quality. The certification of a tool will generally be carried out by an independent, often national, body, against independently set criteria, typically national or international standards. Ideally, the tools used at all development phases — definition, design, coding, testing and validation — plus those used in configuration management, should be subject to certification, but to date only compilers are regularly subject to certification procedures; these are laid down by national certification bodies and they exercise compilers against international standards such as those for Ada^{TM} and Pascal. (Ada is a registered trademark of the US Department of Defense, Ada Joint Program Office).

Conditions: The only preconditions for the use of a certificated tool are its existence and its appropriateness. The preconditions for certifying a tool are the existence of an adequate specification against which a candidate tool can be checked and some agreed process according to which the certification will be carried out.

Major Advantages: Use of a certificated tool gives more confidence in the results it produces, for instance that the object code produced by a compiler has the effect expected of the source code according to the language's definition.

Problems or Disadvantages: No major problems or disadvantages.

Results: The output of a certificated tool can be treated with greater confidence than that of a non-certificated tool.

Assessment: Their use is recommended wherever they are available and appropriate. Certification of a tool is not equivalent to a proof of its correctness in general — particularly in the case of large tools such as compilers. However, certificated tools should be used in preference to non-certificated tools.

Tools: The certification of tools generally requires extensive test data (validation of an Ada compiler requires more than 3000 test programs). The preparation of this test data and the tools to support preparation is generally the responsibility of the certification authority.

References:

1. *Pascal Validation Suite*, UK distributor: BSI Quality Assurance, PO Box 375, Milton Keynes, MK14 6LL.

2. *Ada Validation Suite*, UK distributor: National Computing Centre (NCC), Oxford Rd., Manchester, England.

3. *Ada Evaluation Suite*, BSI Quality Assurance, PO Box 375, Milton Keynes, MK14 6LL.

4. *Pascal Evaluation Suite*, BSI Quality Assurance, PO Box 375, Milton Keynes, MK14 6LL.

3.5   Configuration Management

Aim: Configuration management (CM) aims to ensure the consistency of groups of development deliverables as those deliverables change. It applies to both hardware and software development.

Description: CM is a technique used throughout development. In essence, it requires the recording of the production of every version of every "significant" deliverable and of every relationship between different versions of the different deliverables. The resulting records allow the system developer to determine the effect on other deliverables of a change to one deliverable (especially one of its components). In particular, systems or subsystems can be reliably re-built from consistent sets of component versions.

Conditions: There are no preconditions for effective CM except the establishment of a reliable mechanism for recording the product of a new version of a deliverable and for recording the relationships between deliverables. Thereafter, its implementation is a matter of discipline that can be imposed via tools.

Major Advantages: CM prevents the errors that can arise from uncontrolled changes to deliverables. In particular whilst every care may have been taken in the development of individual components of a system, if the wrong set of component versions is assembled into the final system then the system may be faulty.

Problems or Disadvantages: There can be problems with ensuring the reliability of records. However, all such problems need to be overcome as good CM practices should be regarded as mandatory on any project producing software for a safety critical system. There can be problems when changes are made in the field beyond the control of the CM system.

Results: Consistent sets of deliverables. An ability to determine the knock-on effects of changes to one deliverable.

Related Methods: None.

Assessment: CM should be regarded as a mandatory technique.

Tools: A number of tools are available for CM including MAKE and SCCS on UNIX^{TM} (UNIX is a trademark of Bell Laboratories), CMS and MMS from Digital Equipment Corporation, Lifespan from YARD (Yarrow, England) and CCC from Softool Corporation.

References:

1. *IEEE Standard for Software Configuration Management Plans*. ANSI/IEEE Standard 828–1983.

2. *Configuration Management Practices for Systems, Equipment, Munitions and Computer Programs*. MIL-STD-483.

3. *Software Configuration Management*. J.K. Buckle, Macmillan Press, 1982.

4. *Software Configuration Management*. W.A. Babich, Addison-Wesley, 1986.

5. *Configuration Management Requirements for Defence Equipment*. UK Ministry of Defence Standard 05-57, Issue 1, 1980.

3.6   Development Standards

Aim: To enhance software quality, by using standard approaches to the software development process.

Description: A set of standards are defined covering all aspects of the software development project. This covers such areas as:-

- Project organization.
- Project reviews and inspections.
- Design methods.
- Programming languages and support software.
- Programming style guides.
- Verification and validation methods
- Test methods and levels of testing.
- Documentation standards.
- Version control and configuration management.

Such standards should ensure a more consistent approach to the development. If properly implemented, it should minimise faults by imposing good practice methods (e.g. good programming style) which should also ease subsequent software maintenance.

Conditions: The availability of appropriate standards.

Major Advantages:

- Reduced maintenance problems.
- A basic means for introducing software quality.

Problems or Disadvantages: If the wrong standards are used, it can impede rather than improve the software development process.

Related Methods: Closely related to quality assurance.

Assessment: Essential for safety critical systems. Necessary for implementing a quality assurance program.

Tools: Tool support is available to assist the application of standards during the software development process.

References:

1. *Software Considerations in Airborne Systems and Equipment Certification*, RTCA 178A/EUROCAE ED12A, 1985, Radio Technical Committee for Aeronautics, 1425 K Street N.W., Suite 500, Washington, DC 20005, USA.

2. *Requirements for the Procurement of Safety Critical Software in Defence Equipment*, MOD Interim Defence Standard 00–55, 1989, MOD Directorate of Standardization, 65 Brown St., Glasgow, G2 8EX, Scotland.

3. *Software for Computers in the Safety Systems of Nuclear Power Stations*, IEC 880.

4. *Programmable Electronic Systems in Safety Related Applications: "1 An Introductory Guide"*, Health and Safety Executive, 1987, ISBN 011 8839136, HMSO, PO Box 276, London SW8 5DT.

5. *Programmable Electronic Systems in Safety Related Applications: "2 General Technical Guidelines"*, Health and Safety Executive, 1987, ISBN 011 8839063, HMSO, PO Box 276, London SW8 5DT.

6. *The STARTS Guide*, National Computing Centre (NCC), Oxford Rd., Manchester, England, 1987.

7. *The IT STARTS Developers Guide*, National Computing Centre (NCC), Oxford Rd., Manchester, England, 1989.

8. *Software Engineering Project Standards*, M. Branstad, and D.B. Powell, IEEE Trans. SE-10, Jan 1984.

3.7 Electromagnetic Protection

Aim: To minimize the effects of Electromagnetic Interference (EMI) on the system, by using defensive methods and strategies.

Description: Effective protection against EMI is dependent on the type of disturbances that are expected, on the physical and geometric characteristics of the plant and computer system, and on how the interaction between source and target of the disturbance are linked together. For this reason it is very difficult to establish a general purpose wide frequency range 'fix'. It should be noted that such fixes are very site dependent, and can be more or less effective in different situations. Currently various different categories of what was called EMI have developed.

EMC – Electromagnetic Compatibility.
All the ways that the Electromagnetic Spectrum can influence the system under consideration. These influences are to be considered as abnormal influences.

EMS – Electromagnetic Susceptibility.
The sensitivity that a certain system demonstrates to external electromagnetic pollution.

RFI – Radio Frequency Interference.
The oldest part of this science. A synonym for EMI but referring to the higher frequency range only.

EMP – Electromagnetic Pulse.
Refers to the problem arising from an intense and sudden EM pulse such as one produced by a nuclear explosion.

ESD – Electrostatic Discharge.
Studies the problem of static discharge caused by various sources such as lightning and operator discharge.

Each one of these fields needs different methods to be attenuated. A list of frequently seen fixes as far as compatibility goes is:

- Power line filtering.
- Sound grounding.
- DC power supply buffering.
- Opto-isolation.
- Surge withstand filtering on all incoming input lines.
- High frequency AC filtering.
- Ambient noise spectrum analysis.
- Detailed early study of radiation frequencies, geometries, and mechanisms.
- Use of shielding in a coherent and co-ordinated manner.

Conditions: The main problem is the thorough understanding of the interaction between the environment and system. This is not always easy to achieve. The source can be a wide frequency source and this means that measurements should be performed on the target environment.

Major Advantages: When enough attention is given to the problem many surprises can be eliminated before going into the field.

Problems or Disadvantages: The study of the environment itself is very time consuming, and no two applications have the same disturbance profile (except for repetitive objects like ships, planes etc.).

Related Methods: Can be linked to the qualification phase, and very often feeds back into the design phase.

Assessment: Strongly recommended.

Tools: Recently CAD packages have been developed to automatically calculate frequency profiles for different types of boxes, connection schemes etc. Some actually give the results in graphical form and compare them directly to standards. This reduces the cost of 'testing' a design, and is much less expensive.

References:

1. *Electromagnetic Emission and Susceptibility for the Control of Electronic Interference.* updated 1 April 1987, MIL-STD-461B.

2. *Electromagnetic (Radiated) Environment considerations for design and procurement of electrical and electronic equipment, subsystems, and systems.* MIL-HDBK-235-1A, 18 Dec 1986.

3. *Guide for the installation of electrical equipment to minimize electrical noise inputs to controllers from external sources.* IEEE Std 518–1977.

4. *Guide for surge voltages in low-power circuits.* IEEE Std 587–1980.

5. *Guide for surge withstand capability (SWC) test.* ANSI C.37.90–1974.

6. *Recommended practice for grounding of industrial and commercial power systems.* ANSI C114.1–1973.

7. *Limits and Measurement Methods for Radio Interference from Electronic Data Processing Equipment.* Standard ECMA-47. European Computer Manufacturers' Association.

8. *Limits of radiointerference and leakage current according to CISPR and national regulations.* CISPR, Publication 9–1978.

9. *Specification for radiointerference measuring apparatus and measurement methods.* CISPR Publication 16–1977.

3.8   Formal Development Methods

In this catalogue we have divided software development methods into two broad classes: *formal* methods and *systematic* methods. Formal methods are those that have a mathematical basis. Systematic methods are those that follow traditional good practice — promoting good structuring, modularity, decomposition etc. — but do not have a mathematical basis.

We preface each group of method descriptions with some general statements about the group. We start here with formal development methods. Systematic methods are dealt with in section 3.13.

Aim: The development of software in a way that is based on mathematics. This includes formal design and formal coding techniques.

Description: Formal methods provide a means of developing a description of a system at some stage in its development specification, design or code. The resulting description takes a mathematical form and can be subjected to mathematical analysis to detect various classes of inconsistency or incorrectness. Moreover, the description can in some cases be analysed by machine with a rigour similar to the syntax checking of a source program by a complier, or animated to display various aspects of the behaviour of the system described. Animation can give extra confidence that the system meets the real requirement as well as the specified requirement.

A formal method will generally offer a *notation* (generally some form of discrete mathematics being used), a *technique* for deriving a description in that notation, and various forms of *analysis* for checking a description for different correctness properties.

Conditions: Formal techniques generally require well trained staff. A prerequisite for training is some fluency (or additional training) in discrete mathematics, typically set theory and predicate calculus.

Major Advantages: Formal techniques increase the likelihood that their output (specification, design or code) is correct, i.e. is internally consistent, is consistent with some prior model, possesses certain desirable properties, does not possess certain undesirable properties, and so on. This increased likelihood arises from the possibility of proving these properties mathematically from a mathematically expressed description.

Problems or Disadvantages: Formal techniques suffer from a number of disadvantages. Firstly, they can be more difficult and more time-consuming to use; although these disadvantages may well be outweighed by the criticality of the application. Secondly, they are generally less well supported than, say, the systematic methods in terms of tools; this means that, as with the systematic methods, a great deal of work must be done manually (and hence with a risk of human error) rather than by a computer (and hence with greater confidence). The tool situation is however changing and for more 'popular' methods tool support is starting to become available.

Assessment: Formal methods should be considered for all critical systems. Much research and development work is being undertaken currently in the area and the situation will doubtlessly improve.

References:

1. *An Introduction to Discrete Mathematics and Formal System Specification*, D.C. Ince, Oxford Mathematics and Applied Computing series, ISBN 0-19-859664-2.

3.8.1   CCS – Calculus of Communicating Systems

Aim: CCS is a means for describing and reasoning about the behaviour of systems of concurrent, communicating processes.

Description: Similar to CSP, CCS is a mathematical calculus concerned with the behaviour of systems. The system design is modelled as a network of independent processes operating sequentially or in parallel. Processes can communicate via *ports* (similar to CSP's *channels*), the communication only taking place when both processes are ready. Non-determinism can be modelled. Starting from a high-level abstract description of the entire system (known as a *trace*), it is possible to carry out a stepwise refinement of the system into a composition of communicating processes whose total behaviour is that required of the whole system. Equally, it is possible to work in a bottom up fashion, combining processes and deducing the properties of the resulting system using inference rules related to the composition rules.

Conditions: A working familiarity with discrete mathematics is essential.

Major Advantages: The calculus and the proof theory that comes with it offer the system specifier/designer the opportunity to prove behavioural properties of the system.

Problems or Disadvantages: CCS is weak on the handling of data types though it can be combined with, for example, VDM which gives emphasis to data types, and the problem is overcome in LOTOS which is based on CCS and ACT ONE and which can be expected to form the industrialization route to CCS. It is also weak on handling concrete timing constraints. Like any formal technique its effective use requires some degree of mathematical fluency (though nothing that cannot be acquired by the average university graduate), plus training and experience.

Results: The use of CCS results in a formal specification of a system and, where carried out, proofs of the behavioural characteristics of the system deduced from its composition into processes.

Related Methods: Very similar to CSP.

Assessment: CCS remains, at the time of writing a technique whose application is largely still a research activity, but the industrialization of the method can be expected in the coming years as was the case with VDM. It should be considered at least as a descriptive tool in cases where a system must consist of more than one process.

Tools: There are at the time of writing no support tools available outside the research domain where theorem proving tools are being worked on.

References:

1. *A Calculus of Communicating Systems*. R. Milner, Report number ECS-LFCS-86-7, Laboratory for Foundations of Computer Science, Department of Computer Science, University of Edinburgh, UK.

2. *The specification of complex systems*. B. Cohen, W.T. Harwood, M.I. Jackson. Addison-Wesley, 1986.

3. *ACT ONE: an algebraic specification language with two levels of semantics*, H. Ehrig, W. Frey, H. Hansen, Bericht-Nr. 83-03, Technical University of Berlin, (Berlin 1983).

3.8.2   CSP – Communicating Sequential Processes

Aim: CSP is a technique for the specification of concurrent software systems, i.e. systems of communicating processes operating concurrently.

Description: CSP provides a language for the specification of systems of processes and proof for verifying that the implementation of processes satisfies their specifications (described as a *trace* - permissible sequences of events).

A system is modelled as a network of independent processes. Each process is described in terms of all of its possible behaviours. A system is modelled by composing processes sequentially or in parallel. Processes can communicate (synchronize or exchange data) via channels, the communication only taking place when both processes are ready. The relative timing of events can be modelled.

The theory behind CSP was directly incorporated into the architecture of the Inmos transputer^{TM}, and the occam language allows a CSP-specified system to be directly implemented on a network of transputers.

Conditions: Use of CSP requires some mathematical fluency and specific training in the theory underlying it. Training is available but at the time of writing this comes mostly from university departments and companies specializing in formal methods.

Major Advantages: Its strength is the fact that it handles concurrency explicitly.

Problems or Disadvantages: It is relatively undeveloped compared with VDM. It is weak in its handling of data types and does not explicitly allow handling of concrete timing constraints of the sort often met with in real-time systems. Like any formal technique its effective use requires some degree of mathematical fluency (though nothing that cannot be acquired by the average university graduate), training and experience.

Results: The use of CSP results in a formal model of a system and, when carried out, proofs of the presence or absence of certain behaviours of the system.

Related Methods: The overall approach is similar to CCS . One of the main differences is the different sorts of non-determinism that can be modelled. Work is in progress to combine Z with CSP so that both sequential and concurrent behaviour can be described.

Assessment: CSP remains, at the time of writing a technique whose application is largely still a research activity but the industrialization of the method can be expected in the coming years as was the case with VDM. It should be considered at least as a descriptive tool in cases where a system must consist of more than one process.

Tools: There are at the time of writing no support tools available outside the research domain where tools such as experimental theorem provers are under development. However, the occam language can be used as an execution language. Although designed for the transputer — a processor designed to be used in closely coupled networks — occam can be compiled and executed on single processors such as the VAX^{TM}. (VAX is a registered trademark of Digital Equipment Corporation).

References:

Communicating Sequential Processes. C.A.R. Hoare, Prentice-Hall, 1985.

3.8.3   Finite State Machines

Aim: To define or implement the control structure of a system.

Description: Many systems can be defined in terms of their states, their inputs, and their actions. Thus when in state S_1, on receiving input I a system might carry out action A and move to state S_2. By defining a system's actions for every input in every state we can define a system completely. The resulting model of the system is called a Finite State Machine (FSM). It is often drawn as a so-called *state transition diagram* showing how the system moves from one state to another, or as a matrix in which the dimensions are state and input and the matrix cells contain the action and new state resulting from receipt of the input in the given state.

Where a system is complicated or has a natural structure this can be reflected in a *layered* FSM.

A specification or design expressed as an FSM can be checked for completeness (the system must have an action and new state for every input in every state), for consistency (only one state change is defined for each state/input pair) and reachability (whether or not it is possible to get from one state to another by any sequence of inputs). These are important properties for critical systems and they can be checked. Tools to support these checks are easily written. Algorithms also exist that allow the automatic generation of test cases for verifying an FSM implementation or for animating an FSM model (see the Chow reference below).

Conditions: Use of FSMs requires that the system being developed can be modelled as a black box taking as input a sequence of stimuli. This is common to many process control systems and can in certain circumstances be applied to systems taking asynchronous inputs from different sources.

Major Advantages: FSMs allow certain important properties to be checked mechanically and reliably. They are also simple to work with.

Problems or Disadvantages: FSMs have no major problems or disadvantages.

Results: The use of FSMs results in a model (design or specification) of a system in a semi-formal form which can then be animated. Such a model can then be easily implemented in software and the software tested with test cases that can be generated automatically.

Related Methods: No significant relation to other methods.

Assessment: A simple yet powerful technique that should be considered for event driven systems.

Tools: The SDL (CCITT standard specification language) utilizes extended finite state machines. ASA (Verilog, 150 rue Nicolas Vauquelin , 31081, Toulouse, Cedex, France) permits specification and animation using a finite state machine representation.

Simple tools can be written to carry out analyses of FSMs and for generating test cases using Chow's technique.

References:

1. *Introduction to the theory of finite state machines.* A. Gill, McGraw-Hill, 1962.

2. *A methodology for decomposing system requirements into the processing requirements.* K.G. Salter, Proceedings of the 2nd International Conference on Software Engineering, IEEE, 1976.

3. *Testing software design modelled by finite-state machines.* T.S. Chow, IEEE Trans. Software Engineering, SE-4.

3.8.4   OBJ

Aim: To provide a precise system specification with user feed-back and system validation prior to implementation.

Description: OBJ is an algebraic specification language. Users specify requirements in terms of algebraic equations. The behavioural, or constructive, aspects of the system are specified in terms of operations acting on abstract data types (ADT). An ADT is like an ADA package where the operator behaviour is visible whilst the implementation details are 'hidden'.

An OBJ specification, and subsequent step-wise implementation, is amenable to to the same formal proof techniques as other formal approaches. Moreover, since the constructive aspects of the OBJ specification are machine-executable, it is straightforward to achieve system validation from the specification itself. Execution is essentially the evaluation of a function by equation substitution (re-writing) which continues until specific output value is obtained. This executability allows end-users of the envisaged system to gain a 'view' of the eventual system at the system specification stage without the need to be familiar with the underlying formal specification techniques.

As with all other ADT techniques, OBJ is only applicable to sequential systems, or to sequential aspects of concurrent systems. OBJ has been widely used for the specification of both small- and large-scale industrial applications.

Conditions: A knowledge of algebra to UK 'A' level standard. Around 2-3 days training in formal ADT techniques should be followed by some 2-3 months exposure to the use of OBJ.

Major Advantages:

- Precise specification.
- Capability for rigorous system development.
- Executability, which assists the verification and validation of the specification.

Problems or Disadvantages:

- System designers will need a certain level of mathematical fluency, including familiarity with ADT and stepwise refinement techniques.
- OBJ does not address the concurrency aspects of a system.
- Executability can impose additional constraints on the structure and content of the specification (e.g. speed of execution).

Results: The use of OBJ coupled with rigorous development and test techniques is likely to reduce development risks and costly iterations within the system development and maintenance phases. Documented, real-world applications of OBJ techniques are given in the references.

Related Methods: It can be used as prototyping method. Related to object-oriented design and programming.

Assessment: OBJ is a powerful, yet natural, formal specification language for both large- and small-scale systems developments. Ease of use and executability makes it a serious contender for selection in a formal software development.

Tools:

OBjex, (Gerrard Software, Venture House, Cross St., Macclesfield, SK11 7PG, UK).

An OBJ machine and associated tools are being developed at Stanford Research Institute, USA.

References:

1. *An Introduction to OBJ; A Language for Writing and Testing Specifications.* J.A. Goguen and J. Tardo, Specification of Reliable Software, IEEE Press 1979, reprinted in Software Specification Techniques, N. Gehani, A. McGettrick (eds), Addison-Wesley, 1985.

2. *Algebraic Specification for Practical Software Production.* C. Rattray, Cogan Press, 1987.

3. *An Algebraic Approach to the Standardization and Certification of Graphics Software.* R. Gnatz, Computer Graphics Forum 2(2/3), 1983.

4. *Peacock Reports.* Plessey Research, Roke Manor, Romsey, Hampshire, SO51 0ZN, UK, 1987.

5. *Formal Specification and Design-time Testing.* C. Gerrard, D. Coleman, R. Gallimore, HP-labs, Hewlett Packard, Bristol BS12 6QZ, (to be published in IEEE Transactions).

6. *Implementing Algebraically Specified Abstract Data Types in an Imperative Programming Language.* M. Thomas, Lecture Notes in Computing Science, V.250, Springer Verlag, 1987.

7. *DTI STARTS Guide.* 1987, NCC, Oxford Rd., Manchester, UK.

3.8.5 Petri Nets

Aim: To model relevant aspects of the system behaviour and to assess and possibly improve safety and operational requirements through analysis and re-design.

Description: Petri nets belong to a class of graph theoretic models which are suitable for representing information and control flow in systems exhibiting concurrency and asynchronous behaviour.

A Petri net is a network of places and transitions. The places may be 'marked' or 'unmarked'. A transition is 'enabled' when all the input places to it are marked. When enabled, it is permitted (but not obliged) to 'fire'. If it fires, the input marks are removed, and each output place from the transition is marked instead.

The potential hazards can be represented as particular states (markings) in the model. Extended Petri nets allow timing features of the system to be modelled. Although "classical" Petri nets concentrate on control flow aspects, several extensions have been proposed to incorporate data flow into the model.

Conditions: Mainly applicable to systems dominated by control aspects or concurrent asynchronous activities..

Major Advantages: Method is easy to understand and can provide thorough analysis for small systems. The model is (potentially) executable and this can be used to validate the adequacy of the model. Some classes of fault can be modelled together with their impact on safety (e.g. whether a hazardous state can be reached). Some (rather limited) guidance can be obtained on changes that can be made to the model to make the system tolerant to those faults.

Problems or Disadvantages: The properties of the

whole system are far from obvious from the net's appearance (except very simple nets). The analytic methods work well on small networks or small parts of large networks, but are computationally intractable for large systems. Direct execution is of little help in demonstrating safety as only a small proportion of the overall set of states can be tested.

Results: A model of functional behaviour, an analysis of the safety of the specification, an analysis of the effect of some classes of fault, identification of those faults that threaten system safety.

Related Methods: CSP and CCS are alternative means of modelling and analysing system behaviour. Temporal logic can be used as alternative to demonstrate that the system specification (and design) is safe.

Assessment: Potentially very valuable for small systems or small parts of larger systems.

Tools:

PSItools NET (TOPAS-N), (systems analysis and simulation with Petri nets). PSI, Gesellschaft für Prozesssteuerungs- und Informationssysteme, Gmbh, Heilbronner Strasse 10, D–1000, Berlin 31.

References:

1. *Net Theory and Applications*. W. Brauer (Ed). Lecture Notes in Computer Science, Vol. 84, Springer Verlag, 1980.

2. *Petri Net Theory and Modelling of Systems*. J.L. Peterson. Prentice Hall, 1981.

3. *Safety Analysis using Petri Nets*. N. Leveson and J. Stolzy. Proc. FTCS 15, Ann Arbor Michigan, June 1985. IEEE 1985.

4. *A Tool for Requirements Specification and Analysis of Real-time software based on timed Petri nets*. S. Bologna, F. Pisacane, C Ghezzi, D. Mandrioli, Proc. SAFE-COMP 88, 9–11 Nov. 1988, Fulda, Fed. Rep. of Germany, 1988.

3.8.6   Temporal Logic

Aim: Direct expression of safety and operational requirements and formal demonstration that these properties are preserved in the subsequent development steps.

Description: Standard First Order Predicate Logic contains no concept of time. Temporal logic extends First Order logic by adding modal operators (e.g. 'Henceforth' and 'Eventually'). These operators can be used to qualify assertions about the system. For example, safety properties might be required to hold 'henceforth', whilst other desired system states might be required to be attained 'eventually' from some other initiating state. Temporal formulas are interpreted on sequences of states (behaviours). What constitutes a 'state' depends on the chosen level of description. It can refer to the whole system, a system component or the computer program. Quantified time intervals and constraints are not handled explicitly in temporal logic. Absolute timing has to be handled by creating additional time states as part of the state definition.

Conditions: Skilled personnel to perform proofs in the temporal logic. Best applied to systems whose properties are not dependent on quantified time intervals and constraints.

Major Advantages: Temporal logic provides for a direct expression of the systems behavioural properties. It can be used at different levels of system description. Formal demonstration of consistency between specification levels is possible.

Problems or Disadvantages: There is no machine support for the manipulation of temporal logic specifications or for proving theorems. Hand proofs demand highly skilled personnel. The descriptive nature of the logic is a potential problem since there is little indication of how to proceed to a design.

Results: Formal models at different levels of description. Proofs of consistency between levels.

Related Methods: This method is related to program proving. Petri nets and finite state machines are alternative ways of describing temporal behaviour, but both these approaches are a means of *designing* a system that exhibits the desired behaviour. Temporal logic specifies the behaviour directly.

Assessment: Useful as a descriptive and demonstrative technique for small systems or for small parts of large systems. Applicability to larger systems is strongly dependent on the availability of effective computer-based tools.

Tools: No available commercial tools. Experimental theorem provers exist in universities.

References:

1. *Temporal Logic of Programs*. F. Kroger EATCS Monographs on Computer Science, Vol 8, Springer Verlag, 1987.

2. *Design for Safety using Temporal Logic*. J. Gorski. SAFECOMP 86, Sarlat France, Pergamon Press, October 1986.

3. *Logics for Computer Programming*. D. Gabay. Ellis Horwood.

3.8.7   VDM – Vienna Development Method

Aim: The systematic specification and implementation of sequential programs.

Description: VDM is a mathematically based specification technique and a technique for refining implementations in a way that allows proof of their correctness with respect to the specification.

The specification technique is *model-based* in that the system state is modelled in terms of set-theoretic structures on which are defined *invariants* (predicates), and operations on that state are modelled by specifying their pre- and post-conditions in terms of the system state. Operations can be proved to preserve the system invariants.

The implementation of the specification is done by the *reification* of the system state in terms of data structures in the target language and by refinement of the operations in terms of a program in the target language. Reification and refinement steps give rise to *proof obligations* that establish their correctness. Whether or not these obligations are carried out is a choice made by the designer.

VDM is principally used in the specification stage but can be used in the design and implementation stages leading to source code. It can only be applied to sequential programs or the sequential processes in concurrent systems.

Conditions: The technique requires trained personnel. A prerequisite for training is some fluency (or training) in discrete mathematics (set theory and predicate calculus). A three or four day course is considered adequate for teaching people to read VDM specifications. The development of specifications requires considerably more training and experience.

Major Advantages: VDM offers the system specifier the ability to prove the consistency of a specification and the correctness of a program with respect to that specification. A valuable side-effect is the unambiguity of a VDM specification and the resulting improvement in the communications between user and specifier and between specifier and developer.

System size does not appear a problem for specification and significantly large specifications (in excess of 100 pages) are being specified in VDM including for instance one defining the Portable Common Tool Environment (PCTE).

Problems or Disadvantages: Like any formal technique its effective use requires some degree of mathematical fluency (though nothing that cannot be acquired by the average university graduate), training and experience. Large systems have now been specified in VDM but modularity is not a strong point. Proofs have to be done manually and tool support is only just beginning to appear (e.g. syntax-directed editors, proof checkers, specification management systems). VDM does not permit the modelling of concurrent systems but work is underway within the ESPRIT RAISE project to improve the handling of concurrency and also to improve modularization facilities for large specifications.

Results: Rigorous program specifications whose consistency can be proved as necessary. Implemented programs whose correctness with respect to their specification can be proved.

Related Methods: VDM is closely related to Z especially in the implementation phase. VDM implementation has strong similarities to the program refinement techniques of Dijkstra and Gries (see the references).

Assessment: VDM is an established formally-based technique for which tutorial literature and training courses are available. A number of software engineering companies offer consultancy in VDM and use it to develop systems for clients. It is recommended especially for the specification of sequential programs. Standardization of the syntax of VDM is underway by the BSI in the UK.

Tools: At the time of writing some support tools are appearing, generally syntax-directed editors, syntax checkers, pretty-printers, etc. Proof checkers are being developed. The *SpecBox* tool (Adelard, 28 Rhondda Grove, London, E3 5AP) checks syntax and type usage within a VDM specification. It also generates cross references and has options for pretty-printing. The tool *me too* (see the reference below) has been developed at the University of Stirling, UK, and allows the execution of specifications in a subset of VDM. The ability to execute a specification is an important facility in the development of any system. Although experimentation with such a model cannot (like direct execution of code) prove the correctness of a specification, it can help to increase confidence in it.

References:

1. *Software development — a rigorous approach.* C.B. Jones. Prentice-Hall, 1980.

2. *Formal specification and software development.* D. Bjørner and C.B. Jones. Prentice-Hall, 1982.

3. *Systematic software development using VDM.* C.B. Jones. Prentice-Hall, 1986.

4. *The specification of complex systems.* B. Cohen, W.T. Harwood, and M.I. Jackson. Addison-Wesley, 1986.

5. *VDM 87: A formal method at work.* Lecture Notes in Computer Science. Vol. 252, Springer-Verlag, 1987.

6. *The me too method of software design* P. Henderson and C. Minkowitz, Technical Report TR.14, Department of Computing Science, University of Stirling, 1984.

7. *A discipline of programming.* E.W. Dijkstra, Prentice-Hall, 1976.

8. *The science of programming.* D. Gries, Springer-Verlag, 1981.

3.8.8   Z

Aim: Z is a specification language notation for sequential systems and a design technique that allows the developer to proceed from a Z specification to executable algorithms in a way that allows proof of their correctness with respect to the specification.

Z is principally used in the specification stage but a method has been devised to go from specification into a design and an implementation. It is best suited to the development of data oriented, sequential systems.

Description: Like VDM, the specification technique is *model-based* in that the system state is modelled in terms of set-theoretic structures on which are defined *invariants* (predicates), and operations on that state are modelled by specifying their pre- and post-conditions in terms of the system state. Operations can be proved to preserve the system invariants thereby demonstrating their consistency. The formal part of a specification is divided into *schemas* which allow the structuring of specifications through refinement.

Typically, a Z specification is a mixture of formal Z and informal explanatory text in natural language. (Formal text on its own can be too terse for easy reading and often its purpose needs to be explained, while the informal natural language can easily become vague and imprecise.)

Unlike VDM, Z is a notation rather than a complete method. However an associated method (called B) has been developed which can be used in conjunction with Z. The B method is based on the principle of step-wise refinement.

Conditions: A working knowledge of discrete mathematics is necessary for using Z. However, as with most formal methods (e.g. VDM, CSP and CCS described elsewhere in this book) sufficient theory can generally be taught to university graduates in about one week to allow them to read a Z specification.

Major Advantages: A Z specification — like any formal specification — provides more precise descriptions of an intended system than natural language. A Z specification therefore improves the communication of ideas between specifier and specifier and between specifier and user, and reduces the possibility of misunderstanding between them.

Z has, through the schema mechanism, good facilities for structuring a specification and for reuse of existing parts of specifications.

There are a number of case study specifications available in the literature.

Problems or Disadvantages: Like all formal methods, Z requires its user to be familiar with discrete mathematics. Up to one week's training is regarded as necessary to attain reading level familiarity. A further two week's training and several months' exposure to real applications are generally considered necessary before the method can be successfully applied.

Currently there is no notation for modelling concurrency in systems and no facility for handling timing. The method has not reached the same degree of industrialization and standardization as VDM.

Results: The use of Z results in a specification expressed in a formal language with natural language annotation.

Related Methods: Z has strong similarities in its approach to VDM.

Assessment: Z has shown itself to be a powerful specification notation for large systems, notably (to date) the specification of the programmer's interface to the IBM Customer Information Control System (see Hayes below) and the specification of the public tool interface of the ASPECT integrated project support environment. Whilst it is not as industrialized as VDM it should be considered for the specification of sequential systems. Commercial training is available.

Tools: Toolsets to support Z are under development at the time of writing within the UK Alvey Programme (FORSITE project) and by commercial organizations. A research tool has been developed to support the B method.

References:

1. *The Z Notation - A Reference Manual.* J.M. Spivey, Prentice Hall, 1988.

2. *Specification Case Studies.* Edited by I. Hayes, Prentice-Hall, 1987.

3. *Specification of the UNIX filestore.* C. Morgan and B. Sufrin. IEEE Transactions on Software Engineering, SE-10, 2 March 1984.

3.9 Formally Designed Hardware

Aim: To prove that the hardware design meets its specification.

Description: The basic approach uses the same formal proof methods used for formally designed software. Starting from a mathematically formal specification, the design is transformed by a series of step-wise refinements to a logic circuit design.

Conditions: Currently applied to the central microprocessor chip (in particular the VIPER 1 and VIPER 1A processor chips) rather than the supporting circuits. In order to apply mathematical proof methods, the initial design must be kept simple. This tends to preclude the use of complex instructions (e.g. multiply and divide) or concurrency (such as the use of interrupts).

Major Advantages:

- Precisely specified hardware behaviour.
- High probability that the chip design meets its specification.
- Can be integrated into a general formal development framework for both hardware and software.
- Can be designed to simplify the task of software verification.

Problems or Disadvantages:

- Unspecified behaviour is still possible if there are chip fabrication faults or random hardware failures (some form of hardware redundancy is still needed to detect such failures).
- Unproven support chips and interfaces to plant sensors and actuators limit the level of confidence in the whole

hardware subsystem.

- Unproven support software (e.g. compilers / linkers) limits the level of confidence in the correctness of the system.

- The computer chips tend to be slow, partly because of the limited instruction set, and partly due to the use of well-established chip fabrication methods.

Related Methods: Similar in concept to formal software development. Must be used in conjunction with hardware fault detection and failure management techniques such as cross-checking processor pairs, N out of M voting, or fault-secure design.

An alternative approach to enhancing hardware reliability is to utilize a subset of the hardware instructions of a standard microprocessor. Standard microprocessors have the benefit of extensive field usage for both the hardware and the supporting software.

Results: Improved confidence that the computing hardware system meets its specification.

Tools: Formal specification and proof tools such as OBJ, LCF/LSM etc.

Assessment: A relatively new technique, best applied in a context where *all* components are formally proven. The reliability enhancement achievable with unproven supporting hardware and software is debatable.

Should be considered as an alternative to a standard microprocessor system for relatively simple applications without severe time response requirements.

References:

1. *VIPER Symposium Report*, Safety Net, Issue 1, June 1988, Viper Technologies, PO Box 79, Worcester WR1

2PX, England.

2. *VIPER for Railway Signalling - South Australia*, Safety Net, Issue 4, Mar 1989, Viper Technologies, PO Box 79, Worcester WR1 2PX, England.

3.10   Information Hiding

Aim: To increase the reliability and maintainability of software.

Description: Data that is globally accessible to all software components can be accidentally or incorrectly modified by any of these components. Any changes to these data structures may require detailed examination of the code and extensive modifications.

Information hiding is a general approach for minimizing these difficulties. The key data structures are 'hidden' and can only be manipulated through a defined set of access procedures. This allows the internal structures to be modified or further procedures to be added without affecting the functional behaviour of the remaining software. For example, a name directory might have access procedures *Insert, Delete* and *Find*. The access procedures and internal data structures could be re-written (e.g. to use a different look-up method or to store the names on a hard disk) without affecting the logical behaviour of the remaining software using these procedures.

This concept of an *abstract data type* is directly supported in a number of programming languages, but the basic principle can be applied whatever programming language is used.

Conditions: While all software functions could in principle be implemented in this way, there may be practical constraints (e.g. timing or memory space). It is therefore necessary to identify those functions which will be subject to change over the lifetime of the system. This information should form part of the initial software requirements specification, allowing the software designer to implement information hiding for the most changeable software functions.

Major Advantages: Reduced risk of data corruption faults, simpler design, reduced maintenance effort and reduction in maintainance-induced faults.

Problems or Disadvantages: The use of access procedures can increase execution times. This overhead can be eliminated in some programming languages (e.g. Ada) which have facilities for in-line insertion of procedure code.

Related Methods: Closely related to object-oriented programming and design.

Results: More reliable and maintainable software

Tools: Programming languages providing modularization or object-oriented facilities e.g. Ada, Modula 2, C++, Smalltalk.

Assessment: Highly recommended for all types of software system.

References:

1. *Software Engineering: Planning for Change*, D.A. Lamb, Prentice Hall, 1988.

2. *On the Design and Development of Program Families*, D.L. Parnas, IEEE Trans. SE–2, Mar. 1976.

3.11   Quality Assurance

Aim: To ensure that pre-determined quality control activities are carried out throughout development.

Description: Throughout the development of hardware or software systems various actions will be carried out to check the quality of the work. These actions are collectively referred to as 'quality control' and include the various techniques (or parts of them) described throughout the methods directory. Quality control activities must themselves be planned, monitored and controlled. This activity is called 'quality assurance'. It is concerned with providing a framework for quality control and ways of checking that it is indeed being done.

National and international standards exist for quality assurance — see the references. We can describe the common theme as follows.

Quality assurance should be an activity that takes place independently of the development group. Responsibility for quality assurance and quality control must be clearly allocated to individuals. Guidance and regulations on quality control should be available to staff — in the form of a Quality Manual for instance. Development projects should identify what quality control activities they will undertake during development (perhaps in a quality plan). Detailed records of quality control records should be kept, especially of records of faults found and of the actions taken to correct faults. Development projects should be regularly audited to ensure that the planned quality control activities are taking place and that the records are being kept.

Conditions: The conditions for good quality assurance are described in the relevant standards but we can identify three as being important: adequately defined and measurable quality goals, defined standards for all aspects of development and an independent quality function in the development organization.

Major Advantages: Quality assurance does not itself improve the quality of a system directly, but it is the way in which we can check that enough quality control is done to ensure a certain level of quality in the delivered system.

Problems or Disadvantages: There is an initial cost in setting up the quality assurance function, preparing standards and procedures and training staff. These costs should be saved over a number of developments since, if done properly, it will help to ensure that errors are detected as early as possible and hence as cheaply as possible.

Results: Quality assurance activities result in the audit records for development projects. Quality control activities result in audit records for development activities and in better products of those activities. The final result is a better system.

Related Methods: Quality assurance is applied throughout development and governs all quality control activities. Any method that includes reviewing, checking, proving or auditing can come within its scope.

Assessment: A quality assurance mechanism should be considered mandatory for safety-related systems. Ideally it should conform to a standard such as ISO 9001.

Tools: The day-to-day quality assurance activities typically involve book-keeping which can be supported by tools only to a limited extent. Database systems and some project management support can assist in recording and tracking quality control activities, especially the correction of faults. Tools can be used to perform code quality checks and assist in the testing process (LOGISCOPE, Verilog, 150 rue Nicolas Vauquelin, 31081, Tolouse, France and Testbed Liverpool Data Research Associates, Liverpool Innovation Centre, 131 Mount Pleasant, Liverpool L3 5TS, England).

References:

1. *Quality systems — General introduction and guidelines.*
   ISO 9000–1987.

2. *Quality systems — Model for quality assurance in design/-
   development, production, installation and servicing.*
   ISO 9001–1987.

3. *Specification for design/development, production, installation and servicing.* BS5750: Part 1: 1987, British
   Standards Institution, UK.
   *Note that BS 5750: Part 1: 1987 and ISO 9000–1987
   are the same.*

4. *NATO requirements for an industrial quality control system.* AQAP-1, NATO.

5. *NATO software quality control system requirements.*
   AQAP-13, NATO.

6. *Standard for Software Quality Assurance Plans.*
   ANSI/IEEE Std 730–1984.

7. *Guide for Software Quality Assurance Planning.*
   ANSI/IEEE Std 983–1986.

3.12   Safe Language Subsets

Aim: To reduce the probability of introducing programming faults and increase the probability of detecting any remaining faults.

Description: The language is examined to identify programming constructs which are either error-prone or difficult to analyse (e.g. using static analysis methods). A 'safe' language subset is then defined which excludes these constructs.

Conditions: The language selected should have well-defined semantics, and contain features to aid analysis and avoid errors (e.g. strongly typed, good control structures etc). Languages such as Pascal and Ada Pascal, safe subset Ada, safe subset would be good candidates, while less strict languages such as Fortran and C Fortran, safe subset C, safe subset are less amenable to this approach.

Major Advantages: Reduced risk of residual faults.

Problems or Disadvantages: The subset is less powerful and could be more inconvenient to use, and the programs could be more lengthy.

Related Methods: Usually used in association with coding quality guidelines (e.g. use of meaningful names, use of global data, limits on code complexity, size of modules etc). Related to analysable software and complexity metrics.

Results: Simple, analysable software which should contain fewer faults and be easier to maintain.

Tools: Documented safe language subsets, tools to check adherence to the language subset.

Assessment: Highly recommended for safety-related software.

References:

1. *Spade Pascal* , Program Validation Ltd., 34 Basset Crescent East, Southampton SO2 3FL, England.

2. *SPARK Ada subset*, Program Validation Ltd., 34 Basset Crescent East, Southampton SO2 3FL, England.

3.13   Stress Reduction

Aim: To ensure that under all normal operational circumstances both hardware components and software activity are operated well below their maximum stress levels.

Description: The reliability of hardware components is increased by operating all components at levels guaranteed by the design of the system to be well below the maximum component specification ratings. The reliability of software is improved by ensuring, through system design, that the maximum software activity and memory space requirements are, respectively, well below the time needed to complete the software activity and well above the maximum memory requirement of the software.

Hardware examples include:

- components operating at a fraction of their maximum junction temperature;
- components operating at a fraction of their maximum speed;
- components operating well within their environmental specification;
- communications systems operating at a fraction of their rated bandwidth.

Software examples include:

- under all operational conditions software idle time is significant;
- under all operational conditions only a fraction of the available memory space or segment space is utilized.

Conditions: System performance criteria must be obtainable (this may include cost).

Major Advantages: Actual and calculated reliability of hardware components will be improved with a resultant increase in availability. Software enhancements will be more easily achieved due to available memory space and time.

Problems or Disadvantages: Over-size, under-rated components may cause design constraints or make design impractical. Additional cost of hardware components due to software stress reduction requirements may be unacceptable.

Related Methods: Stress reduction forms an integral relationship with the system requirement specification. The use of certificated hardware components and aspects of Software Quality Assurance are also related.

Assessment: It is good industrial design practice to minimize stress both in hardware and software. On safety critical systems, additional stress reduction methods should always be taken when practical.

Tools: None.

References:

1. *Reliability Prediction of Electronic Equipment.* MIL-HDBK-217, DOD, USA.

2. *Reliability of Electronic Components.* BS 4200, British Standards Institute.

3.14 Strongly Typed Programming Languages

Aim: Reduce the probability of faults by using a language which permits a high level of checking by the compiler.

Description: Such languages usually allow user-defined data types to be defined from the basic language data types (such as INTEGER, REAL). These types can then be used in exactly the same way as the basic types, but strict checks are imposed to ensure the correct type is used. These checks are imposed over the whole program, even if this is built from separately compiled units. The checks also ensure that the number and the type of procedure arguments match even when referenced from separately compiled modules.

Strongly typed languages also support other aspects of good software engineering practice such as easily analysable control structures (e.g. IF .. THEN .. ELSE, DO .. WHILE, etc) which lead to well-structured programs.

Typical examples of strongly typed languages are Pascal, Ada and Modula 2.

Conditions: To be properly used, a disciplined top-down approach to software design and coding is needed, where the major data types and procedure interfaces are defined prior to detailed development.

Major Advantages:

- Reduced numbers of programming faults.
- More maintainable software.

Problems or Disadvantages: No major disadvantages.

Related Methods: Closely related to other software engineering approaches such as structured programming, con-

figuration management, analysable programs, static analysis etc.

Results: Well structured, consistent software.

Tools: An appropriate compiler and support enviroment.

Assessment: Highly Recommended.

References:

1. *Reference Manual for the Ada Programming Language*, ANSI/MIL-STD-815A 1983.

2. *In search of Effective Diversity: a Six Language Study of Fault-Tolerant Flight Control Software*, A. Avizienis, M.R. Lyu, W. Schutz, The Eighteenth International Symposium on Fault Tolerant Computing, Tokyo, Japan, 27–30 June 1988, IEEE Computer Society Press, ISBN 0-8186-0867-6.

3.15   Structured Programming

Aim: To design and implement the program in a way which makes practical the analysis of the program. This analysis should be capable of discovering all significant program behaviour.

Description: The program should contain the minimum of structural complexity. Complicated branching should be avoided. Loop constraints and branching should (where possible) be simply related to input parameters. The program should be divided into appropriately small modules, and the interaction of these modules should be explicit. Features of the programming language which encourage the above approach should be used in preference to other features which are (allegedly) more efficient, except where efficiency takes absolute priority (e.g. some safety-critical systems).

Conditions: The design notation and programming language must express the intention of the program in a clear way. An inflexible notation or language will obscure the intention.

Major Advantages: A high degree of confidence in the correctness of the unexecuted program can be obtained informally. The executing program can more readily be analysed by a walkthrough.

Problems or Disadvantages: It is not difficult to find problems which are inherently complex and whose solution cannot be clearly expressed using available notations. Real-time applications introduce requirements to do with timing and exception handling, and often give rise to programs which do not have a simple control flow. Timing constraints on program execution demand specific (and often non-structured) programming measures.

Results: Clearer programs, higher confidence in correctness, easier to maintain, easier to test.

Related Methods: Aids program analysing and testing. Formal program proving.

Assessment: Good programming practice. Should be used wherever possible.

Tools: Some tools for program design. Some automatic measurement of the degree of program structuring by 'static analysis' tools:–

- Logiscope^{TM} (Verilog, rue Nicolas Vauquelin, 31081 Toulouse, Cedex, France).
- Spade^{TM} (Program Validation Ltd., 34 Basset Crescent East, Southampton SO2 3FL, England).
- Malpas^{TM} (Rex Thompson Partners, Newnhams, West Street, Farnham GU9 7EQ, England).
- Testbed^{TM} (Liverpool Data Research Associates, Liverpool Innovation Centre, 131 Mount Pleasant, Liverpool L3 5TS, England.

References:

1. *A Discipline of Programming.* E.W. Dijkstra, Englewood Cliffs NJ, Prentice-Hall, 1976.

2. *Assessing a class of software tools.* M.A. Hennell et. al., 7th International Conference on Software Engineering, March 1984, Orlando.

3. *A software tool for top-down programming.* D.C. Ince, Software — Practice and Experience, Vol. 13, No. 8, August 1983.

3.16   Structuring the System according to Criticality

Aim: To reduce the complexity of safety critical software.

Description: The main cause of problems with software is its complexity. The software functions can be classified according to their criticality with respect to safety. The distinction between safety critical and non-safety-critical functions should be preserved and supported by the design. Then, extra effort can be devoted to increase the quality of the safety critical part. The quality of the non-critical part can receive less attention (from the safety perspective). Nevertheless, any arguments related to system safety which refer to the separation have to be supported by the analysis that demonstrates that there is no way system safety can be compromised by faults in the non-critical software (e.g. implementing the critical software in a separate computer and strictly limiting the interface to the rest of the software).

Conditions: Can be applied only if the separation of functions with respect to safety criticality is possible.

Major Advantages: Supports separation of concerns and reduces the complexity of software. Provides for identification of those parts of the software where the main effort should be devoted. Simplifies any subsequent licensing or certification activity.

Problems or Disadvantages: Requires very strong arguments that the separation is fully maintained (this may require extended hardware architecture, increased analytical effort, etc.).

Results: The approach can be used during the system requirements specification and analysis and then during design phases.

Related Methods: The information used in this process is derived from a plant safety analysis (e.g. using System Hazard and Operability Studies, Fault Tree Analysis, and Failure Modes and Effects Analysis methods).

Assessment: Strongly recommended where applicable.

Tools: None known.

References:

1. *Dependability of Critical Computer Systems: Part 2.* E.F. Redmill (ed.), Elsevier Applied Science, 1989.

3.17 Systematic Analysis and Design Methods

Aim: The main aim of systematic analysis and design (SAD) methods is to promote the quality of software development by focussing attention on the early parts of the life-cycle. The methods aim to achieve this through both precise and intuitive procedures and notations (assisted by computers) to identify the existence of requirements and implementation features in a logical order and a structured manner.

Description: A range of SAD methods exist. Some such as SSADM, LBMS are designed for traditional data-processing and transaction processing functions, while others (MASCOT, JSD, Real-Time Yourdon) are more oriented to process-control and real-time applications (which tend to be more safety-critical).

SAD methods are essentially "thought tools" for systematically perceiving and partitioning a problem or system. Their main features are:

- A logical order of thought, breaking a large problem into manageable stages.
- Identification of total system, including the environment as well as the required system.
- Decomposition of data and function in the required system.
- Checklists, i.e. lists of the sort of things that need definition.
- Low intellectual overhead – simple, intuitive, pragmatic.

The supporting notations tend to be precise for identifying problem and system entities (e.g. processes and data flows), but the processing functions performed by these entities tend to be expressed using informal notations. However some methods do make partial use of (mathematically) formal notations (for example JSD makes use of regular expressions; Yourdon, SOM and SDL utilise finite state machines). This precision not only reduces the scope for misunderstanding, it provides scope for automatic processing.

Another benefit of the SAD notations is their visibility, enabling a specification or design to be checked intuitively by a user, against his powerful but unstated knowledge.

Conditions: Adequate training in the method.

Major Advantages

- Provides a systematic approach to the definition of the requirements.
- Good structuring of data and of software components is encouraged.
- Such methods are generally easier to learn than fully formal methods.
- There is support for concurrency in some methods.

Problems and Disadvantages Many of the methods lack a precise notation for specifying what the components actually do or mean, as opposed to their structure. Data flow interactions may not be properly specified. It may be difficult to demonstrate that successive levels of design implement their specification.

Results: The method results in specifications and designs. With tools static and dynamic checks are possible, and in some cases program code (or partial programs) can be generated.

Related Methods: Related to finite state machines, structured programming, prototyping. Can be used in conjunction with mathematically formal methods.

Assessment: An improvement on traditional, less prescriptive development methods. Structured Analysis is recommended as a means of avoiding oversights in the initial requirements. With tool support, structured analysis and design methods are valuable as a means of avoiding clerical errors and inconsistencies. Systematic methods should be combined with formal specification approaches on systems of high criticality.

Tools: Most of these methods are supported by tools providing some or all of the following:–

- Display editing, to establish tables and diagrams in a machine-readable form.
- Syntax checks.
- Simple consistency checks (e.g. for data type consistency) within and between diagrams.
- A central database or data dictionary to support consistency checking and software administration (e.g. time-stamping, version numbering etc).
- Code generation or the generation of code 'templates'.

References

1. *Structured development for real-time systems* (three volumes) P.T. Ward and S.J. Mellor. Yourdon Press, 1985.

2. *Essential Systems Analysis.* St. M.McMenamin, F. Palmer, Yourdon Inc., 1984, N.Y.

3. *The Use of Structured Methods in the Development of Large Software-Based Avionic Systems.* D.J. Hatley, Proceedings DASC, Baltimore, 1984.

4. *Strategies for Real-Time Systems Specification.* Derek J. Hatley, I.A. Pirbhai, Dorset House Publ. Co., 1987, N.Y.

5. *A Practical Handbook for Software Development*, N.D. Birrell and M.A. Ould, Cambridge University Press, 1987.

6. *CASE-Tools emerge to handle Real-Time Systems.* H. Falk, Computer Design, p. 53-74, Jan 1, 1988.

7. *The STARTS Guide.* National Computer Centre, Manchester, UK, 1987.

8. *The STARTS Purchasers Handbook.* National Computer Centre, Manchester, UK, 1989.

3.17.1 CORE – Controlled Requirements Expression

Aim: To ensure that all the requirements are identified and expressed.

Description: This approach is intended to bridge the gap between the customer/end user and the analyst. It is not mathematically rigorous but aids the communication process – CORE is designed for requirements *expression* rather than specification. The approach is structured and the expression goes through various levels of refinement. The CORE process encourages a wider view of the problem, bringing in a knowledge of the environment in which the system will be used and the differing viewpoints of the various types of user. CORE includes guidelines and tactics for recognising departures from the 'grand design'. Departures can be corrected or explicitly identified and documented. Thus specifications may not be complete, but unresolved problems and high-risk areas are identified and have to be addressed in the subsequent design.

Conditions: Adequate training in the method and notations. Tool support is useful.

Major Advantages: Gives higher-level viewpoints of the system requirements, and hence reduces the chance of producing an inappropriate system.

Problems or Disadvantages: No guarantee of completeness.

Results: Structured requirements description, and documentation of the higher levels of design.

Related Methods: Can be used in conjunction with other analysis and design approaches.

Tools:

- CORE ANALYST (Systems Designers, Camberley, Surrey, UK).
- CORE workbench (British Aerospace, Warton, Lancs, UK).

Assessment: This type of systematic approach to requirements expression is recommended for safety-critical systems (especially for aspects that cannot be expressed formally but are still safety-related e.g. maintenance, security and operator interfaces).

3.17.2 JSD – Jackson System Development

Aim: A development method covering the development of software systems from requirements through to code, with special emphasis on real-time systems.

Description: JSD is a staged development process in which the developer models the real world processes upon which the system functions are to be based, determines the required functions and inserts them into the model, and transforms the resulting specification into one that is realizable in the target environment. It therefore covers the traditional phases of definition, design and implementation but takes a somewhat different view from the traditional methods in not being top-down.

Moreover, it places great emphasis on the early stage of identifying the entities in the real world that are the concern of the system being built and on modelling them and what can happen to them. Once this analysis of the 'real-world' has been done and a model created, the system's required functions are analysed to determine how they can fit into this real-world model. The resulting system model is augmented with structured descriptions of all the processes in the model and the whole is then transformed into programs that will operate in the target software and hardware environment.

Conditions: As with all methods, development staff must be trained in its use and a project using the method should have a core of staff who are already familiar with its use.

Major Advantages: The method's emphasis on the modelling of the real-world first rather than functionality makes the resulting design less liable to disruption in the event of changes of requirement. The resulting system structure is also closely related to the structure of the problem to be solved.

The method is designed to be applied to real-time systems

consisting of concurrent processes.

Problems or Disadvantages: No major problems or disadvantages.

Results: Fully documented design and code.

Related Methods: No significant relationship to other methods but has similarities to MASCOT.

Assessment: JSD should be considered for real-time systems where concurrency can be allowed and where great formality is not called for.

Tools: Some support tools are available, generally for the clerical activities of the method:–

Speedbuilder, PDF (Michael Jackson Systems Ltd, 22 Portland St London W1N 5AF)

References:

1. *An Overview of JSD*. J.R. Cameron, IEEE Trans. SE-12, no. 2 Feb, 1986.

2. *System Development*. M. Jackson, Prentice-Hall, 1983.

3.17.3   MASCOT

Aim: The design and implementation of real-time systems.

Description: MASCOT (Modular Approach to Software Construction, Operation and Test) is a design method supported by a programming system. It is a systematic method of expressing the structure of a real-time system in a way that is independent of the target hardware or implementation language. It imposes a disciplined approach to design that yields a highly modular structure, ensuring a close correspondence between the functional elements in the design and the constructional elements appearing in system integration. A system is designed in terms of a network of concurrent processes that communicate through channels. Channels can be either *pools* of fixed data or *queues* (pipelines of data). Control of access to channels is defined independently of the processes in terms of *access mechanisms* that also enforce scheduling rules on the processes. Recent versions of MASCOT have been designed with Ada implementation in mind.

MASCOT supports an acceptance strategy based on the test and verification of single modules and larger collections of functionally related modules. A MASCOT implementation is intended to be built upon a MASCOT *kernel* — a set of scheduling primitives that underline the implementation and support the access mechanisms.

Conditions: MASCOT is appropriate for real-time systems in situations where a design in terms of concurrent processes is acceptable. Ideally it should be used in conjunction with a MASCOT kernel.

Major Advantages: One strength is that it handles concurrency explicitly.

Problems or Disadvantages: No major problems or disadvantages.

Results: An operational system with supporting design documentation.

Related Methods: MASCOT has affinities with JSD in the construction of a system model in terms of communicating processes.

Assessment: MASCOT should be considered for the development of real-time systems where concurrency has to and can be used.

Tools: Tool support is available in the form of support environments such as :–

- *Modular Approach to Software Construction operation and Test*, Def Stan 00-17, Issue 1, 1985, UK Ministry of Defence.
- PERSPECTIVE^{TM}, CONTEXT^{TM} from Systems Designers, Camberley, Surrey, UK.
- MASCOT 700 from Software Sciences/Ferranti Computer Systems Ltd.

References:

MASCOT 3 User Guide. MASCOT Users Forum, RSRE, Malvern, England, 1987.

3.17.4 Object-Oriented Design and Programming

Aim: To reduce development and maintenance costs and enhance reliability, through the production of more maintainable and re-usable software.

Description: Object-oriented design is a modular decomposition technique based on an object-centred model rather than the more conventional data/procedure model used in functional decomposition methods.

Object-oriented design (OOD) evolved from developments in object-oriented languages (OOLs) such as Smalltalk and Simula 67. It is not a complete methodology like JSD and it does not necessarily depend on the use of an object-oriented language. Some of the concepts are applicable to any software implementation language, but conventional languages do restrict the range of options available in implementing an object-oriented design. Following problem definition and the identification of a solution strategy, an object-oriented design is formalized by:-

- identifying objects and their attributes,
- identifying the operations on these objects,
- establishing interfaces among the objects and operations.
- implementing the operations.

Object oriented-languages simplify the process of implementing objects and their operations. The main features of an object-oriented language are:-

Object definition. Objects are a form of *abstract data type*. An object is defined by a set of data variables which define its state, together with a description of the valid operations to be performed on this data. These operations are sometimes referred to as *methods* or *features*. This approach eases maintenance by more closely mapping real-world entities on to objects in the computer model.

Encapsulation. Access to the data is only possible through the pre-determined operations included in the object definition. Direct access to the state variables is prohibited which guarantees the integrity of the representation. This is an implementation of the Parnas *information hiding* principle which enables the internal structure of an object to be re-organised without affecting other objects in the system.

Classes and instances. A *class* is a description of the properties of one or more similar objects and corresponds to the definition of an *abstract data type*. An *instance* of a class is a single object described by that particular class. A class can be viewed as a 'template' which is used in creating an instance of an object.

Sub-classes and Inheritance. This is an important feature of object-oriented design and programming which enhances the re-usability of software. This feature enables a new class to be defined as a modification of an existing class. The new class inherits the operations of its parent class to which new operations can be added (or the existing operations can be re-defined).

Message passing. In some languages objects communicate through *messages*. While this is convenient for supporting information exchange between objects, it is not essential.

Conditions: The approach is readily learned, but mastery requires some experience. Designers accustomed to functional decomposition methods have to learn a new way of thinking. Also, identifying the objects and their operations is not always straightforward especially when designing for re-usability.

Major Advantages:

- Promotes software re-use, which should lead to reduced costs and increased productivity.

- It is likely to produce better and more reliable software, due to the use of information hiding principles and the minimization of coupling between modules.

- The abstract data type and information hiding principles can be applied to any software design.

- Good language support.

Problems or Disadvantages:

- Object identification is sometimes difficult. It is often unclear what objects and operations most easily map on to the problem domain.

- Many object-oriented languages run more slowly than conventional languages since they are implemented using an interpreter and features such as the creation of classes and instances, inheritance and message passing involve run-time overheads. Compiled languages are available, although there is some compromise between execution speed and language capabilities.

Results: Re-usable software with less complexity and better reliability.

Related Methods: Jackson Structured Design uses similar concepts. The specification and implementation language OBJ also makes use of the concept of abstract data types.

Assessment: Object-oriented design principles are recommended as one possible option for the design of safety-related systems. Object-oriented design and prototyping are also recommended for the construction of prototypes. However care must be taken in the use of object-oriented languages for target applications which are time-critical. In these cases, the use of languages such as ADA and C++ is recommended which incorporate some of the concepts, but remain relatively efficient at run-time.

Tools: Languages which can be considered to be Object-Oriented are:– SIMULA, SMALLTALK, Loops, Objective C, C++, Eiffel, Object Pascal.

Languages such as ADA and Modula 2 support information hiding principles through packages. ADA also supports classes to some degree through the use of generic type definitions.

References:

Object-Oriented Systems Analysis. S. Shlaer and S.J. Mellor, Yourdon Press, 1988.

Extensions and Foundations of Object-Oriented Programming. J.A. Goguen, J. Meseguer, Object Oriented Programming Workshop, SIGPLAN Notices, ACM, Vol 21, No. 10, October 1986.

Object Oriented Programming: An evolutionary Approach. B.J. Cox, Addison Wesley, 1988.

On the Criteria for Decomposing Systems into Modules. D.L. Parnas, CACM, Dec 1972.

Object-Oriented Software Construction. B. Meyer, Prentice Hall, 1988.

Software Engineering with ADA. G. Booch, Addison-Wesley, 1982.

Software Components with ADA - Structures, Tools and Subsystems. G. Booch, The Benjamin/Cummins Publ. Corp. Inc., Menlo Park, California, 1987.

Smalltalk 80: The Language and its Implementation. A. Goldberg and D. Robsonn, Addison-Wesley, Reading, Mass, 1983.

CommonLoops: Merging Common Lisp and Object-oriented Programming. D.G. Bobrow, J. Kahn, G. Kiczales, L. Masinter, M. Stefik and F. Zdybel, Xerox Palo Alto Research Centre, ISL-85-8, August 1985.

SIMULA an Algol-based Simulation Language. O.J. Dahl and K. Nygaard, Comm. ACM, Vol 9. pp 671-678, 1966.

The C++ Programming Language. B. Stroustrup, Addison-Wesley, Reading, Massachusetts, 1986.

3.17.5   Real-Time Yourdon

Aim: This aims to be a complete software development method consisting of specification and design techniques oriented towards the development of real-time systems.

Description: The development scheme underlying this technique assumes a three phase evolution of a system being developed. The first phase involves the building of an 'essential model', one that describes the behaviour required by the system. The second involves the building of an implementation model which describes the structures and mechanisms that, when implemented, embody the required behaviour. The third phase involves the actual building of the system in hardware and software. The three phases correspond roughly to the traditional definition, design and implementation phases but lay greater emphasis on the fact that at each stage the developer is engaged in a modelling activity.

The essential model is in two parts: the environmental model containing a definition of the boundary between the system and its environment and a description of the external events to which the system must respond, and the behavioural model which contains schemas defining the transformation the system carries out in response to events and a definition of the data the system must hold in order to respond.

The implementation model also divides into sub-models: in this case covering the allocation of individual processes to processors and of the decomposition of the processes into modules.

To capture these models the technique combines a number of well-known techniques: data-flow diagrams, structured English, state transition diagrams and Petri Nets.

Additionally the method contains techniques for simulating a proposed system design either on paper or mechanically from the models that are drawn up.

Conditions: Real-time Yourdon explicitly handles systems of concurrent processes. This might be considered too risky for high integrity systems and the use of the entire method would be unnecessary. The techniques involved are not too formal in the mathematical sense and do not require extensive training.

Major Advantages: The method explicitly handles systems of concurrent processes.

Problems or Disadvantages: The method is systematic but still informal.

Results: The method results in specifications and designs.

Related Methods: The method is an amalgam of well-known techniques — see above — and is a development of Structured Analysis and Structured Design.

Assessment: Worth considering for real-time systems without a level of criticality that demands more formal approaches.

Tools: Tool systems are available for the entire method and for the individual modelling techniques. They are mainly diagramming and clerical tools with some consistency checking:–

- MacCadd (Logica plc, 64 Newman St., London W1A 4SE).
- Excelerator/RTS (Excelerator Software Products Ltd, Prince Edward St., Berkhamsted, HP4 3AY, UK).
- Yourdon Analyst/Designer Toolkit (Yourdon Inc, 1501 Broadway, New York, NY 10036, USA; Yourdon International, 15/17 Ridgmount St., London WC1E 7AH).

References:

Structured development for real-time systems (three volumes) P.T. Ward and S.J. Mellor. Yourdon Press, 1985.

3.17.6   SADT – Structured Analysis and Design Technique

Aim: To model and identify, in a diagrammatical form using information flows, the decision making processes and the management tasks associated with a complex system.

Description: In SADT^{TM} (SofTech, Inc.) the concept of an Activity-Factor Diagram plays a central role. An A/F diagram consists of activities grouped in so called 'action boxes'. Each action box has a unique name, and is linked to other action boxes by *factor relations* (drawn as arrows) which are also given unique names. Each action box can be hierarchically decomposed into subsidiary action boxes and relations. There are four types of factors: inputs, controls, mechanisms and outputs:–

Input indicated by an arrow that enters an action box at the left hand side. Inputs can represent material or immaterial things and they are suitable for manipulation by one or more activities in an action box.

Control are typically instructions, procedures, choice criteria and so on. Controls guide the execution of an activity and they are shown by arrows entering the top side of an action box.

Mechanism is a resource such as personnel, organisational units or equipment, that is needed for an activity to perform its task.

Output can denote anything that an activity produces, and it is pictured by an arrow leaving an action box at the right hand side.

When activities are strongly related to each other by many factor relations then it is perhaps better to consider these activities as an indivisible group that is contained in one action box which does not lend itself to further detailing of its content. The guiding principle for grouping of activities into action boxes is that the resulting boxes are coupled pairwise by only a few factors.

The model hierarchy of A/F diagrams is pursued until a further detailing of the action boxes is meaningless. This stage is reached when the activities within the boxes are inseparable or when further detailing of the action boxes falls outside the scope of the system analysis.

Conditions: A thorough understanding of the system to be analysed.

Major advantages: SADT permits the analysis of any activity in a complex system. The resultant analysis may be viewed at differing levels of detail.

Problems or disadvantages:

- When activities are grouped together in action boxes, care must be taken that no boxes overlap, i.e. one activity is assigned to more than one box.
- The unique naming convention makes it difficult for different activities to share potentially common mechanisms (e.g action boxes containing pre-programmed functions).

Related Methods: Related to systems theory.

Assessment: SADT is a good analysis tool for existing systems, and can also be used in the design specification of systems. Recommended in assessment of safety critical functions and relationships in complex systems (plant level).

Tools: Not known.

References:

1. *Structured Analysis for Requirements Definition*, D.T.

Ross, K.E. Schoman Jr., IEEE Trans. Software Eng., Vol. SE-3, 1, 1977, 6-15.

2. *Structured Analysis (SA): A language for communicating ideas*, D.T. Ross, IEEE Trans. Software Eng., Vol. SE-3, 1, 1977, 16-34.

3. *Applications and extensions of SADT*, D.T. Ross, Computer, April 1985, 25-34.

4. *Structured Analysis and Design Technique - Application on Safety Systems*, W. Heins, Risk Assessment and Control Courseware, Module B1, chapter 11. 1989, Delft University of Technology, Safety Science Group, P.O. Box 5050, 2600 GB Delft, Netherlands.

5. *SPECIF-X: A tool for CASE*, M. Lissandre, 9th IEEE Conference Software Engineering, Monterey, Cal., March 1987.

3.17.7   SDL - Specification and Description Language

Aim: To be a standard language for the specification and design of telecommunication switching systems.

Description: SDL contains a language part and a set of user guidelines. The language is based upon the Extended Finite State Machine (EFSM). The system to be made is decomposed into processes which are represented as EFSM processes communicating with other processes and the environment. The process always is in a specified state, and when some input reaches it, a defined action is executed and a transition is made to a new state, where a new input is awaited. In this Other diagrams are used that help to gain an understanding of the system behaviour. The SDL methodology has an abstract data part based upon ACT1 for data description.

Conditions: Some training is required.

Major advantages: It is strong on control, which is important for telecommunication systems. It can be tested and verified, having a formal model defined. It handles concurrency.

Problems or disadvantages: The methodology is not fully developed, and it seems that most users use SDL as a kind of flowchart. The data part is not easy to use, and it may be difficult to comprehend what the system actually does.

Results: Some documented design. Its guidelines do not cover the full documentation of a project, as it is meant to be used in conjunction a documentation system.

Related Methods: SDL is based upon the EFSM. It is similar to SOM.

Assessment: Should be considered as a possible option for a specification and design methodology, especially for telecommunications systems.

Tools: Many tools exist for editing SDL diagrams. Some tools produce executable code from the process charts.

References:

1. *SDL '87. State of the Art and Future Trends* (eds.) R. Saracco, P.A.J. Tilanus; Elsevier Science Publishers, 1987.

2. *CCITT Recommendation Z.100*. CCITT, Geneva, 1987.

3.17.8 SOM - Systems development by an Object-oriented Methodology

Aim: A development language and methodology covering the development of systems consisting of software and hardware from requirements to implementation, with special emphasis on real time systems.

Description: SOM contains a language part and a methodology part, guiding the user in his system definition. The language contains many different expression forms, whereof the most important is the state-transition model that is founded upon the extended finite state machine (EFSM). The system is decomposed into functional processes, that are each an EFSM process communicating with other processes and the environment. The process is always in a specified state, and when some input reaches it, a defined action is performed and the EFSM makes a transition to a new state, where it waits for new input. Other types of diagram help to get an understanding of the system behaviour, and abstract data type definitions are used for data description.

In SOM one makes firstly a functional model, and then an architectural model of the system's software and hardware. This division helps to reuse functional processes, library functions and support systems.

Conditions: Some training is required.

Major advantages: SOM can express temporal behaviour and data handling aspects. Concurrency can be readily expressed in terms of interacting state machines. SOM models can easily be tested and verified. Models can be developed to describe both the implemented system and the behaviour of external systems.

Problems or disadvantages: A relatively new method. Currently used in Norway. At the time of writing, documentation on the method is only available in Norwegian.

Results: System documentation, that are 1) specifications and 2) program code.

Related Methods: SOM is based upon the EFSM, which can be expressed in terms of a mathematical notation SBC that is an extension of CCS, and which models the EFSM specifications formally. It is related to SDL and possible to use within a framework that uses a restricted SDL. The document handling part of the methodology is adopted from SADT.

Assessment: Used in Norwegian industry, should be considered as an option for a development methodology.

Tools: There is one tool available: DASOM (Data-Assisted SOM). At the current stage of development, some expression forms in SOM are not yet supported.

References:

1. *SOM - Language overview.* G. Hasnes, Elab report, Elab, Trondheim, Norway, 1987.

2. *Lecture notes in Software Engineering.* R. Brök, Elab report, 1988.

3.17.9 Synchronous Data Flow Specification Languages

Aim: A structured specification and implementation expressed in terms of parallel processes and data flow.

Description: Automatic control systems are intrinsically parallel. They are generally specified in terms of boolean equations, gates, switch networks, block diagrams or transfer functions.

In earlier discrete logic and analog functions, the implementation mirrored the inherent parallelism of the required functions. In computer-based implementations, these functions had to be represented in sequential terms in order to express them in terms of conventional programming languages.

Synchronous data flow languages considers each program as a function whose outputs are defined by equations on the inputs, and respond instantaneously to those inputs. Logic that requires a memory of a prior system state can be expressed by including outputs from a function as part of the input to the function.

Conditions: Some training in the technique is required

Major Advantages: It simplifies the implementation of functions that have intrinsically parallel modes of operation.

Strong typing eliminates incorrect usage of data and functions. Explicit initialisation of the initial input values a potential source of failure.

Problems or Disadvantages: Tool support is only just beginning to appear. Not particularly appropriate for numerically intensive computations (e.g. numerical algorithms). Such functions may have to be treated as 'black boxes' which are implemented in conventional programming languages.

Assessment: A relatively new method. Should be considered as a possible approach for the implementation of concurrent real-time control systems.

Tools: LUSTRE, SAGA.

References:

1. *LUSTRE: a declarative language for programming synchronous systems.* P. Caspi, E. Pilaud, N. Halbachs, J.A. Plaice, Fourteenth ACM Symposium on 'Principles of Programming Languages'.

2. *SAGA: a software development environment for dependability automatic controls.* J.L. Bergerand, E. Pilaud, SAFECOMP 88, 9–11 Nov. 1988, Fulda, Fed. Rep. of Germany, Pergamon Press, 1988.

Chapter 4

Fault Detection

4.1 Analysable Programs

4.2 Design for Testability (Hardware)

4.3 Design for Testability (Software)

4.4 Flow Analysis

4.5 Inspections and Walkthroughs

4.6 Performance Modelling

4.7 Program Proving

4.8 Prototyping

4.9 Symbolic Execution

4.10 Testing

    4.10.1 Tests Based on Random Data
    4.10.2 Tests Based on Realistic Data
    4.10.3 Tests Based on the Specification
    4.10.4 Tests Based on Software Structure
    4.10.5 Back-to-Back Testing

4.11 Test Adequacy Measures

4.12 Verification and Validation

4.1   Analysable Programs

Aim: To design a program in a way that program analysis is easily feasible. The program behaviour must be testable completely on the basis of the analysis.

Description: The intention is to produce programs which are easy to analyse using static analysis methods. In order to achieve this, the rules of structured programming should be followed, e.g.:–

- The module control flow should be composed of structured constructs – sequence, iteration (e.g. DO and WHILE) and selection (e.g. IF .. THEN .. ELSE ..).
- The modules should be small.
- The number of possible paths through the module is small.
- The individual program parts have to be designed so that they are decoupled as far as possible.
- The relation between the input parameters and output parameters should be as simple as possible.
- Complex calculations should not be used as the basis of branching and loop decisions.
- Branch and loop decisions should be simply related to the module input parameters.
- Boundaries between different types of mappings shall be simple.

Conditions: The problem to be solved has to be fairly simple. For example, all problems that can be described by decision tables can be completely treated in this way.

Major Advantages: Code can be verified by testing. Program proving is facilitated.

Problems or Disadvantages: May not qualify for complex problems, such as simulation of plant behaviour, Fast Fourier Transform or mathematically demanding tasks. No real-time features may be explicitly built in; the problem has to be so simple that all time constraints are fulfilled without any specific measures.

Related Methods: Complementary to program analysis and program proving.

Assessment: Recommended for wherever possible. Essential if the verification process makes use of static program analysis techniques.

Tools:

- Logiscope^{TM} (Verilog, rue Nicolas Vauquelin, 31081 Toulouse, Cedex, France).
- Spade^{TM} (Program Validation Ltd., 34 Basset Crescent East, Southampton SO2 3FL, England).
- Malpas^{TM} (Rex Thompson Partners, Newnhams, West Street, Farnham GU9 7EQ, England).
- Testbed^{TM} (Liverpool Data Research Associates, Liverpool Innovation Centre, 131 Mount Pleasant, Liverpool L3 5TS, England.

References:

1. *MALPAS: Verification of a Safety Critical System.* J.T. Webb and D.J. Mannering, SARSS 87, Nov. 1987, Altrincham, England, Elsevier Applied Science, ISBN 1-85166-167-0, 1987.

2. *Verification - The Practical Problems.* B.J.T. Webb and D.J. Mannering, SARSS 87, Nov. 1987, Altrincham, England, Elsevier Applied Science, ISBN 1-85166-167-0, 1987.

3. *An Experience in Design and Validation of Software for a Reactor Protection System.* S. Bologna, E. de Agostino et. al., IFAC Workshop, SAFECOMP '79, Stuttgart, 16–18 May 1979, Pergamon Press, 1979.

4. *A Computerized Protection System for a Fast Research Reactor.* S. Bologna, E. de Agostino et. al., IEEE 1979 Symposium on Nuclear Power Systems, San Francisco, 17–19 Oct. 1979, Proceedings published in IEEE Trans Nuclear Science, Vol. NS-27, No. 1, Feb. 1980.

5. *Programanalysis — A Method for the Verification of Software for the Control of a Nuclear Reactor.* W.D. Ehrenberger, et. al., 2nd International Conference on Software Engineering, San Francisco, 13–15 Oct. 1976.

4.2   Design for Testability (Hardware)

Aim: To enable all hardware components to be fully tested both on and off line.

Description: Designing for off-line testability requires the careful positioning of extra test points and sockets. Full functional testing of a hardware assembly may then be carried out off-line.

Designing for on-line testability requires that feed-back and monitoring are both present, and that there are no unexercised and therefore untested components in the system. Testability should extend not only to the final switching or output elements but also include feedback on the effect of the output action.

Conditions: Off-line testability requires the availability of adequate functional test equipment. A prerequisite for on-line testability is feedback monitoring and redundancy if correct and safe control is to be maintained.

Major Advantages: On-line testing of hardware components is vital if system reliability in a redundant system is to be maintained. Latent faults are detected and reported, fault detection enhanced and maintainability improved.

Problems or Disadvantages: To enable the elimination of untested components, such as diodes, requires advanced design techniques. The addition of testing and monitoring circuits may reduce operational reliability.

Results: Safety function availability increased, fault detection improved.

Related Methods: Usual to integrate on-line testing into fault tolerant system to detect latent faults.

Assessment: Should be used whenever fault tolerance and redundancy is applied.

Tools: Markov modelling simulators, Fault tree analysers. Functional and ATE. Test Equipment. Signature Analysis.

References:

1. *Design for Testability - A Survey.* Proc. IEEE, Vol. 71, No. 1, 1983.

2. *Designing ATE Testable PCB's.* (Hardware and Software together in Microprocessor Systems. C.G. Mckay, Factron-Schlumberger, 1981 IEE Symposium, IEE, Savoy Place London WC2R OBL.

3. *Built-in self test techniques.* E.J. McCluskey, IEEE Design and Test of Computers, Vol. 2, No. 2, April 1985.

4. *Gracefully degradable processor arrays.* J.A.B. Fortes et. al., IEEE Trans. Comp., Vol. C-34, No. 11, Nov. 1985.

5. *Design Principles for processor maintainability in real-time systems.* H.Y. Chang and J.M. Scanlon, In 1969 Fall Joint Computer Conference, IEEE

6. *A Unified Built-in Self Test Scheme: UBIST.* M. Nicolaidis, FTCS 18, Tokyo, 27–30 July 1988, Tokyo, IEEE, 1988.

4.3 Design for Testability (Software)

Aim: To make software amenable to thorough testing.

Description: Throughout the design of software, attention should be paid to the question of selecting and executing a set of test data which will give a high degree of confidence that the software is correct. At higher design levels, this might involve selecting algorithms and data structures with low intrinsic complexity (e.g. avoid the use of dynamic pointers). At lower levels, it may involve such things as simplifying or minimizing the number of paths through the program. Or, a variable which is technically private to a procedure might be made externally available for monitoring.

Conditions: There are no particular limits to this approach.

Major Advantages: Testing is currently the most widely practised and understood technique. The more thorough the testing can be, the better.

Problems or Disadvantages: Extra code to monitor behaviour, therefore less than optimal algorithms and data structures. Module isolation is lessened when private variables are exposed for monitoring. Test probes in real-time code inevitably alter its execution time. Problem of removing extra code after testing.

Results: Increased confidence in the correctness of the software.

Related Methods: Some aspects are common to general minimization of complexity.

Assessment: Strongly recommended.

Tools: Test harnesses, coverage analysers and automatic test data generators are all available, but the basic design activity is subjective.

References:

1. *The theory and practice of functional testing.* W.E. Howden, IEEE Software, Vol. 2, No. 5, Sep. 1985.

2. *Practical priorities in system testing.* N.H. Petschenik, IEEE Software, Vol. 2, No. 5, Sep. 1985

3. *SEES — a Software Testing Environment Support System.* N. Roussopoulos et. al., IEEE Trans. Soft. Eng., Vol. SE-11, No. 4, April 1985.

4. *SADAT — an automated testing tool.* U. Voges et. al., IEEE Trans. Soft. Eng., Vol. SE-6, No. 3, May 1980.

4.4 Flow Analysis

Aim: To detect poor and potentially incorrect program structures.

Description:

Flow analysis identifies suspect areas of code that do not follow good programming practice. There are two main types of flow analysis:–

- *Control Flow Analysis* where the the program is analysed to form a directed graph which can be analysed for:–

  - Inaccessible code e.g. an unconditional jump round a block of code leaves an island of unreachable code.
  - 'Knotted' code – well structured code has a control graph which is *reducible* by successive graph reductions to a single node. Poorly structured code can only be reduced to a 'knot' composed of several nodes.

- *Data Flow Analysis* combines the information obtained from the control flow analysis with information about which variables are read or written in different portions of code. The analysis can check for:–

  - Variables that are read before they are written. This is very likely to be an error, and is certainly bad programming practice.
  - Variables that are written more than once without being read. This could indicate omitted code.
  - Variables that are written but never read. This could indicate redundant code.

There is an extension of data flow analysis known as information flow analysis, where the actual data flows (both within and between procedures) are compared with the design intent. This is normally implemented with a computerized tool where the intended data flows are defined using a structured comment that can be read by the tool.

Conditions: Availability of the source code. For information flow analysis, a knowledge of the intended data flows is needed.

Major Advantages:

- Simple to apply.
- Readily automated.

Problems or Disadvantages:

- Results require some interpretation. The identified anomalies may not be faults.
- Sometimes difficult to deal with 'aliasing' where different variables are associated with the same locations in memory (however this is a poor programming practice).

Related Methods: Complementary to inspection methods. Structured programming will reduce the number of anomalies detected.

Assessment: Recommended especially if there is suitable tool support.

Tools:

- Logiscope^{TM} (Verilog, rue Nicolas Vauquelin, 31081 Toulouse, Cedex, France).
- Malpas^{TM} (Rex Thompson Partners, Newnhams, West Street, Farnham GU9 7EQ, England).

- Testbed^{TM} (Liverpool Data Research Associates, Liverpool Innovation Centre, 131 Mount Pleasant, Liverpool L3 5TS, England.
- RXVP80^{TM} (General Research Corporation, Santa Barbara, Cal., USA).
- Spade^{TM} (Program Validation Ltd., 34 Basset Crescent East, Southampton SO2 3FL, England).

References:

1. *RXVP80 – The Verification and Validation System for FORTRAN. User's Manual.* General Research Corporation, Santa Barbara, California, USA.

2. *Information flow and data-flow of while programs,* J.F. Bergeretti and B.A. Carré, ACM Trans. on Prog. Lang. and Syst., 1985, 7, 37-61.

3. *Lecture Notes on Program Validation,* B.A. Carré, 1982, Dept. of Electronics and and Information Engineering, University of Southampton, Southampton, England.

4.5   Inspections and Walkthroughs

Aim: To detect errors in some product of the development process as soon and as economically as possible.

Description: A number of different forms of inspection have been described in the literature and these are then often adapted by organizations to fit their own requirements and practices. The two best known are Fagan Inspections and Structured Walkthroughs. In this summary we describe the common features.

An inspection starts with the distribution of the item to be inspected - say a specification, a set of test data or some code - to all those who are required to take part. Each participant is required to analyse the item on his own - perhaps from a particular viewpoint - and to look for possible problems. The inspection itself is a meeting of those people at which the item is jointly analysed with the aim of trying to find as many errors as possible. During the inspection all errors found are recorded. No attempt is made to correct the errors at the inspection itself. At some time in the future a check is made that the errors have indeed been corrected. If the errors were significantly serious or the changes required were significantly large then a new inspection might be held to check the piece of work again.

Conditions: Little training is required but new participants need to understand the rules under which the inspection is held. These are generally designed to ensure that the inspection is as cost-effective as possible. An organization using inspections is advised to draw up its own rules of conduct for inspections - Yourdon's book is especially good for guidance in this respect.

Major Advantages: Errors are discovered as soon as possible with the result that expensive re-working later on is avoided. Because different people look for errors from different viewpoints more errors might be found than if a single person were to inspect the item; detection rates of up to 80% have been claimed. Statistics are generated which can be used to monitor the inspection efficiency.

Problems or Disadvantages: If an inspection is not properly controlled it can waste time. It is the responsibility of the inspection's chairperson or moderator to ensure that this does not happen.

Results: An inspection results in written record of the errors found in the item being inspected. Subsequently the item is corrected and the record updated to show that the errors found were corrected. This information can be subsequently analysed to monitor and/or improve the inspection process.

Related Methods: Inspections can be used in conjunction with any method that produces something that can be inspected.

Assessment: A very effective method of finding errors that should be adopted throughout the software development process.

Tools: The method requires no tools but document configuration control and distribution tools can help.

References:

1. *Structured Walkthroughs* E. Yourdon, Yourdon Press, 1979.

2. *Design and code inspections to reduce errors in program development.*
   M.E. Fagan, IBM Systems Journal, 15, 3, 1976.

3. *Handbook of Walkthroughs, Inspections and Technical Reviews: Evaluating Programs, Projects and Products*, D.P. Freedman and G.M. Weinberg, Third ed. Little, Brown and Co., Boston, MA, USA, 1982.

4.6   Performance Modelling

Aim: To ensure that the working capacity of the system is sufficient to meet the specified requirements.

Description: The requirements specification includes throughput and response requirements for specific functions, perhaps combined with constraints on the use of total system resources (e.g. processor capacity). The proposed system design is compared against the stated requirements by:–

- defining a model of the system processes, and their interactions,
- identifying the use of resources by each process (e.g. processor time, communications bandwidth, storage devices etc),
- identifying the distribution of demands placed upon the system under average and worst-case conditions,
- computing the mean and worst-case throughput and response times for the individual system functions.

For simple systems, an analytic solution may be possible. For more complex systems, some form of simulation is required to obtain accurate results.

Before detailed modelling, a simpler 'resource budget' check can be used which summates the resource requirements of all the processes. If the requirements exceed designed system capacity, the design is infeasible. Even if the design passes this check, performance modelling may show that excessive delays and response times occur due to resource starvation. To avoid this situation engineers often design systems to use some fraction (e.g. 50%) of the total resources so that the probability of resource starvation is reduced.

Conditions: An accurate model of the system functions and their interactions is required. The analysis requires an estimate of the resource usage and elapsed time of each system function. These estimates can be obtained in several ways: comparison with existing systems, prototyping and benchmarking of time-critical systems, queueing theory and simulation applied to subsystems.

Major Advantages: Early identification of infeasible requirements.

Problems or Disadvantages: It may be difficult to obtain accurate estimates of function times and resource usage. The use of simulation demands that the system is modelled in detail, and this detail recorded in some simulation language. This may be a considerable programming exercise with its own reliability problems.

Results: Information concerning the behaviour of the final system, but always subject to the effort taken and the accuracy of the modelling technique.

Related Methods: May be used as a form of prototyping. Good systems analysis and design is a prerequisite for performance modelling.

Assessment: Extremely valuable provided the modelling limitations mentioned above are recognised.

Tools: GPSS, SIMSCRIPT, Network II, Simula, Smalltalk, Concurrent Pascal, Modula 2, Ada.

References:

1. *The development of GPSS*. G. Gordon, ACM Sigplan Notices, Vol. 13, 1978.

2. *SIMSCRIPT*. Proc. Conf. on Simulation measurement and modelling of computer systems. University of Colorado, Aug. 1979.

3. *Queues.* D.R. Cox and W.L. Smith, Methuen monographs on applied probability and statistics, Methuen 1961.

4.7 Program Proving

Aim: To check whether software fulfils its intended function.

Description: For every designed software control structure a specific predicate logic pre-condition and post-assertion can be defined which should be true for any program execution. Formal proof methods can then be applied to the code to check whether the assertion conditions are met.

Conditions: The technique requires specialist effort and automated proving aids are also needed for programs of any size. It has to be applied in the context of a formal program development where the code has been designed for program verification and proof methods.

Major Advantages: A formal proof of correctness is possible. The assertions can also be used as the basis for test condition generators and run-time error traps to detect failures caused by faults in compilers, computer hardware or other software components.

Problems or Disadvantages: Very time-consuming. Difficult to apply to large software systems. Only applicable to sequential programs — concurrent program interactions are not catered for. Must verify that the assertions are correct.

Results: Very high quality software. A set of assertions that can be used to check that the software has been correctly transformed to executable code.

Related Methods: The technique is used in formal program specification and development.

Assessment: Should be used for the key software components of a safety-critical system.

Tools: Proving aids, such as IPE (a prototype interactive proof editor) and the concurrency workbench (NFCS, Computer Science Dept, Edinburgh University).

References:

1. *Assigning meaning to programs.* R.W. Floyd, Proc. Symp. Applied Math Vol. 19, Math Aspects of Computer Science (J.T. Schwarz ed.).

2. *The axiomatic basis of computer programming.* C.A.R. Hoare, ACM Trans. Vol. 12, Oct. 1969.

3. *A Discipline of Programming.* E.W. Dijkstra, Prentice Hall 1976.

4. *The rigorous development of a system version control program.* I.D. Cottam, IEEE Trans. Soft. Eng., Vol. SE-10, No. 2, March 1984.

5. *Tentative steps toward a development method for interfering programs.* C.B. Jones, ACM Trans. Prog. Lang. and Systems, Vol. 5, No. 4, October 1983.

6. *Proving liveness properties of concurrent programs.* S. Owicki and L. Lamport. ACM Transactions on Programming Languages and Systems, Vol. 4, no. 3, 1982.

7. *Verifiication of concurrent programs: a temporal proof system.* Z. Manna and A. Pnueli, STAN-CS-83-967. Dept of Computer Sci., Stanford University, 1983.

4.8   Prototyping

Aim: To check the feasibility of implementing the system against the given constraints. To communicate the specifier's interpretation of the system to the customer, in order to locate misunderstandings.

Description: A sub-set of system functions, constraints and performance requirements are selected. A prototype is built using high-level interactive tools. At this stage, constraints such as the target computer, implementation language, program size, maintainability, reliability and availability need not be considered. The prototype is evaluated against the customer's criteria and the system requirements may be modified in the light of this evaluation.

Conditions: Existence of a system specification containing quantified objectives and an explicit definition of the system function. Adequate channels of communication between the customer and the supplier.

Major Advantages:

- Better communication with the customer.
- Early detection of problems (contradictions, omissions, impossibility).
- Can check the feasibility of new ideas or techniques.

Problems or Disadvantages:

- Unnecessary and expensive if the problem is well understood.
- Needs tools for quick implementation.
- There is a danger that the prototype may be used in the final product.

Results: Better agreement between customer and supplier on the system functionality. More confidence that the system is feasible.

Related Methods: Formal specification methods, performance simulation. May be tested in conjunction with a simulation of the plant connected to the system.

Assessment: Valuable if the system requirements are uncertain or the requirements need strict validation.

Tools: Executable specification languages, powerful descriptive languages with a rich support environment (e.g. OBJ, metoo, EPROS, SETL, Prolog, LISP, Loops, Smalltalk, APL), supporting tools (graphics, windows), simulator tools.

References:

1. Proc. Working Conference on Prototyping, Namur, Oct 1983, Budde et al, Springer-Verlag, 1984.

2. *Using an executable specification language for an information system.* S. Urban et. al., IEEE Trans. Soft. Eng., Vol. SE-11, No. 7, July 1985.

3. *The application of some artificial intelligence tools and techniques in software engineering.* D.C. Ince, M. Woodman, S. Hekmatpour, in "Software Engineering: the Decade of Change", D.C. Ince (ed.), Peter Peregrinus, 1986.

4. *An operational approach to requirements specification for embedded systems.* P. Zave, IEEE Trans. SE-8, no. 3, 1982.

4.9   Symbolic Execution

Aim: To show the agreement between the source code and the specification.

Description: The progam is executed substituting the left hand side by the right hand side in all assignments. Conditional branches and loops are translated into Boolean expressions. The final result is a symbolic expression for each program variable. This can be checked against the expected expression.

Conditions: Formal specifications (with assertions) are necessary to check the results.

Major advantages:

- No input data values are needed.
- The results can be used in program proving.

Problems or disadvantages:

- The result will consist of algebraic expressions which easily get very bulky and difficult to interpret.
- Difficult to analyse loops with variable length.
- For most programs the number of possible symbolic expressions is excessively large.
- Unlikely to detect missing paths.

Results: Output variables as algebraic or logic functions of the input variables.

Related Methods: Program correctness proofs can be performed over symbolic execution trees. Can assist in determining *loop invariant* assertions needed for program proofs.

Assessment: Recommended for safety critical software providing the number of paths is small and there is good tool support.

Tools:

- Spade^{TM} (Program Validation Ltd., 34 Basset Crescent East, Southampton SO2 3FL, England).
- Malpas^{TM} (Rex Thompson Partners, Newnhams, West Street, Farnham GU9 7EQ, England).

References:

1. *Formal Program Verification using Symbolic Execution.* R.B. Dannenberg and G.W. Ernst, IEEE Trans Software Engineering, Vol. SE-8, No. 1, 1982.

2. *Symbolic Execution and Software Testing*, J.C. King, Comm. ACM, Vol. 19, No. 7, 1976.

4.10 Testing

Aim: To detect faults by detecting deviations from expected behaviour when the software is executed.

Description: The software is executed in a controlled and systematic way as a means of demonstrating the presence of the required behaviour and the absence of unwanted behaviour. Using some test data selection strategy, a sequence of input values are applied to the software and the output values are evaluated against some test criterion.

A variety of test data selection strategies can be employed:-

- Random, which takes no account of the software function.
- Realistic, where the input sequences are in some way 'typical'.
- Specification-based, which relies on a description of its external behaviour.
- Structural, where tests are based on the internal software structure.

There are many different forms of test criteria, including:-

- logical correctness
- numerical accuracy
- time response
- throughput
- reliability

Test criteria can also include more qualitative attributes such as ease of use and maintainability.

There are also many different means of implementing such tests, including:-

- Expected output values for specific sequences of input values.
- Set of rules or assertions.
- Comparison with an executable specification.
- Comparison with another program/system.

Tests are applied at different stages of software development:–

- Unit testing. This is applied to individual software modules where the primary test criteria will be logical correctness, accuracy and time response.
- Integration testing. Testing of intermediate software systems. The range of test criteria may expand to cover aspects such as throughput reliability and ease of use. Note that different types of integration are possible. The conventional approach is 'bottom-up' testing starting with individual modules. However 'top-down' testing is also feasible (with the design of suitable lower-level 'stub' routines). This permits top-down design and testing to proceed in parallel.
- System testing and acceptance testing. At this stage, tests should cover all significant attributes of the system e.g.:–
    - response time
    - throughput (e.g. transactions per second)
    - storage capacity
    - reliability
    - availability
    - safety (e.g. fail-safe bias)
    - response to overload (e.g. graceful degradation)
    - ergonomics (e.g. time to learn, speed of use, error rate)
- Regression tests. When software is maintained or modified, regression tests should be applied to ensure that other parts of the software have not been affected by the changes. The test could comprise any of the other tests discussed above.

Conditions: Preferably performed by an independent tester. Testing by the developer is useful, since the developer has an intimate knowledge of the program. However developers are not likely to see the faults in their own code. An independent tester will bring a different viewpoint and is also likely to approach testing with a more positive attitude; finding a fault is regarded as success rather than failure.

Major Advantages: Since it is performed on the executable code, compilation-related and program related faults can be revealed.

Problems or Disadvantages: It is not usually possible to exhaustively test software of any significant size. Testing can therefore only seek to show the *presence* of faults; it cannot demonstrate the complete *absence* of faults.

Related Methods: Makes use of test coverage measures and a range of test strategies. Complementary to static analysis methods.

Results: Documented evidence of software quality. A test suite that can be used for subsequent maintenance. A test suite is an extremely valuable resource and should be maintained under configuration control as an integral part of the software.

Tools: A wide range of tools are available to implement unit and integration testing.

Assessment: Essential for any critical computing system.

References:

1. *The Art of Software Testing*, G. Myers, Wiley & Sons. N.Y, USA, 1979.

2. *Software Testing Techniques.* B. Beizer, Van Nostrand Reinhold, New York, 1983.

3. *Software Validation, Verification, and Testing Technique and Tool Reference Guide*, NBS Special Publication 500-93. U.S. Department of Commerce 1982.

4. *Software Test Documentation*, ANSI/IEEE Std 829, 1983.

5. *Code of Practice for Testing Computer-based Systems*, BS5887, 1980, BSI, UK.

4.10.1 Tests based on Random Data

Aim: To cover test cases not covered by systematic methods. To minimise the effort of test data generation.

Description: A test harness is devised which can generate random test data sets. Various types of random data set can be utilized, such as pure uniform random, random around boundaries, gaussian distributions and random walk distributions where the input data is altered by small random increments. There is some evidence to show that this test data strategy is quite effective, giving similar levels of branch coverage, statement coverage and fault detection efficiency to systematic methods. The main problem is to determine which tests result in a failure, so this technique is mainly used in conjunction with comparison testing.

Conditions: Requires special test harnesses and test data generators. Requires some means of determining the correctness of the resultant output.

Major Advantages:

- No program and specification analysis required.
- A large number of test data sets can be produced with little effort.
- Can be an effective form of testing.
- Not prone to human bias and 'mind-set'. For example, systematic tests may check that outputs are set, but not check whether they are cleared.

Problems or Disadvantages:

- With a large amount of output, failures can be hard to detect without an automated check of 'correctness'.
- Purely random data may be inappropriate (e.g. random text for a compiler), so the right type of randomness has

to be selected.

Results: Increased confidence resulting from the large numbers of test cases applied and the removal of human bias.

Related Methods: Complementary to other test data selection strategies. Only useful if there is an automated means of detecting anomalous behaviour.

Assessment: Recommended if there is some automated means of detecting anomalous or incorrect behaviour (such as diverse programs and comparison testing, or possibly run-time assertion checks).

Tools: Random number generator algorithms and subroutines.

References:

1. *Evaluation of Random Testing*, J.W. Duran, S.C. Ntafos, IEEE Trans SE-10, No. 4, July 1984.

2. *STEM - A Project on Software Test and Evaluation Methods*, P.G. Bishop, *et al*, in Achieving Safety and Reliability with Computer Systems (ed. B.K. Daniels), SARSS 87, Elsevier Applied Science, ISBN 1-85166-167-0, 1987.

3. *PODS - An Experiment in Software Reliability*, P.G. Bishop, *et al*, IEEE Trans. SE–12, No. 9, 1986.

4.10.2   Tests based on Realistic Data

Aim: To detect faults likely to occur under realistic operating conditions.

Description: Test data sequences are selected which are regarded as 'typical'. This can be achieved by recording data on similar systems or by using simulations of the intended plant.

Major Advantages:

- Gives a better indication of the reliability of the system under operational conditions.
- Can reveal faults associated with the original specification.

Problems and Disadvantages:

- Tends to cover a smaller region of the input domain, so it tends to be less effective in detecting faults.
- It is sometimes difficult to define an adequate set of 'typical' operational scenarios.

Related Methods: Complementary to other test data selection strategies.

Tools: Plant simulators, plant monitoring systems.

Assessment: Not particularly effective or appropriate at the early stages of software development. Recommended for system testing and acceptance testing.

References:

1. *The Art of Software Testing*, G. Myers, Wiley & Sons. N.Y, USA, 1979.

2. *Software Testing Techniques.* B. Beizer, Van Nostrand Reinhold, New York, 1983.

3. *Software Validation, Verification, and Testing Technique and Tool Reference Guide*, NBS Special Publication 500-93. U.S. Department of Commerce 1982.

4. *STEM - A Project on Software Test and Evaluation Methods*, P.G. Bishop, *et al*, in Achieving Safety and Reliability with Computer Systems (ed. B.K. Daniels), SARSS 87, Elsevier Applied Science, ISBN 1-85166-167-0, 1987.

4.10.3 Tests based on the Specification

Aim: To check whether there are any faults in the program which cause deviations from the specified behaviour of the software.

Description: This approach can be applied at any level of software integration provided there is an appropriate specification of its behaviour. The test sets should be selected to cover all aspects of the specification, and designed to maximize the probability of fault detection. There are a number of techniques which can be employed:–

Equivalence Partitioning The input domain of the program is divided into a number of input equivalent classes (inputs which produce a similar result). Tests are selected from each equivalence class, this limits the number of test cases but may not reveal problems close to domain boundaries.

Boundary-value analysis The input domain of the program is divided into a number of input classes. The tests should cover the boundaries and extremes of the classes. Choices often include maximum, minimum, slightly overrange and trivial values. The tests check that the boundaries in the input domain of the specification coincide with those in the program.

Cause-effect graphing *Causes* (distinct input conditions) and *Effects* (output conditions) in the specification are identified and translated into a Boolean graph. This graph is used to derive test cases in a systematic manner representing combinations of conditions. This has the advantage of providing a different perspective on the specification through its translation into a Boolean graph, and can highlight unwanted side effects. However boundary values are not examined in detail.

Conditions: Precise, unambiguous and well-structured specifications at all levels from software modules to the entire program.

Major Advantages: Tests are directly related to required behaviour.

Problems or Disadvantages:

- Translation of the specification into tests is currently labour intensive.
- It may be impossible to design tests which cover all cases due to the large number of potential combinations.

Related Methods: Complementary to other test data selection strategies.

Assessment: Essential part of an overall test strategy.

References:

1. *Functional Program Testing*, W.E. Howden, IEEE Trans. SE-6, 1980.

2. *A Domain Strategy for Program Testing*, L.J. White and E.A. Cohen, IEEE Trans. SE-6, 1980.

3. *The Art of Software Testing*, G. Myers, Wiley & Sons. N.Y, USA, 1979.

4. *Toward a Theory of Test Data Selection*, J. Goodenough and S.L. Gerhart, IEEE Trans SE-3, 1977.

4.10.4 Tests based on Software Structure

Aim: To apply tests which exercise certain subsets of the program structure.

Description: Based on an analysis of the program, a set of input data is chosen such that a large fraction of selected program elements are exercised. The 'program elements' exercised can vary depending on the level of rigour required:–

Statements. This is the least rigorous test since it is possible to execute all code statements without exercising both branches from a conditional statement.

Branches. Both sides of every branch should be checked. This may be impractical for some type of defensive code.

Compound Conditions. Every condition in a compound conditional branch (i.e. linked by AND / OR) is exercised.

LCASJ. A *Linear Code Sequence and Jump* is any linear sequence of code statements (including conditional jumps) terminated by a jump. Many potential sub-paths will be infeasible due to constraints on the input data imposed by the execution of earlier code.

Data Flow. The execution paths are selected on the basis of data usage. For example a path where the same variable is both written and read.

Call Graph. A program is composed of subroutines which may be invoked from other subroutines. The call graph is the tree of subroutine invocations in the program. Tests are designed to cover all invocations in the tree.

Entire Path. Execute all possible paths ('threads of execution') through the code. Complete testing is normally infeasible due to the very large number of potential paths.

Conditions: A program analysis to determine the input conditions which exercise different branches in the code.

Major advantages:

- A systematic form of testing.
- Test generation can be automated in some cases.

Problems or disadvantages:

- Sometimes difficult to determine the test data required.
- Complete test coverage is usually difficult as some program elements are inaccessible due to input data constraints.
- Does not find missing code, logic or data flow.

Results: Quantifiable test coverage of the program structure.

Related Methods: Complementary to other types of testing. Relies on program structure analysis.

Assessment: Essential as part of an overall test strategy for critical systems. Coverage of all branch conditions and the call graph is recommended for safety critical programs.

Tools: Some tools exist to assist program analysis, the application of tests and measurement of test coverage.

References:

1. *A Comparison of Some Structural Testing Strategies*, S.C. Ntafos, IEEE Transactions SE-14, No. 6, June 1988.

2. *The Path Prefix Software Testing Strategy*, R.E. Prather, J.P. Myers, IEEE Transactions SE-13, No. 7, July 1987.

3. *Experience with Path Analysis and Testing of Programs*, M.R. Woodward, M.A. Hennel and D. Hedley, IEEE Trans SE-6, 1980.

4. *Reliability of the Path-based Testing Strategy*, Howden W.E., IEEE Trans SE-2, 1976.

5. *A Data Flow Analysis Approach to Program Testing*, P.M. Hermann, Australian Computing Journal, Vol. 8, No. 3, Nov 1976.

6. *An Applicable Family of Data Flow Testing Criteria*, P.G. Frankl, E.J. Weyuker, IEEE Trans. SE-14, No, 10, Oct 1988.

4.10.5   Back-to-Back Testing

Aim: To detect test failures by comparing the output of two or more programs implemented to the same specification.

Description: This technique is also known as comparison testing. The same input data is applied to two or more program versions and their outputs are compared to detect anomalies. Any test data selection strategy can be used for this type of testing, although random testing is well suited to this approach.

The programs used in the tests can be:–

- Diverse programs intended for use in a diversely programmed final system.
- A main program and a secondary program intended solely for testing.
- An operational program and an executable form of the original specification (written in a very high level language unsuitable for real-time application).

In the last two cases, it is possible to apply comparison testing down to the subroutine level since the secondary 'program' is intended solely for testing, so there is no requirement to preserve design diversity for run-time operation.

Conditions: A source of test data and suitable test harnesses are required.

Major Advantages:

- Permits a large number of tests to be made with little effort.
- Rapid fault detection.

Problems and Disadvantages:

- Requires the construction of least one secondary program, although this may available as part of the overall development.
- Discrepancies must still be analysed manually to determine which program is at fault (it is not sufficient to assume that the majority is always correct, or that one program is 'golden').

Tools: None known.

Assessment: Recommended if two or more programs are to be produced as part of the normal development process. To be considered as a possible test option if only one version is intended to be produced.

References:

1. *STEM – A Project on Software Test and Evaluation Methods*, P.G. Bishop, *et al*, in Achieving Safety and Reliability with Computer Systems (ed. B.K. Daniels), SARSS 87, Elsevier Applied Science, ISBN 1-85166-167-0, 1987.

2. *PODS – An Experiment in Software Reliability*, P.G. Bishop, *et al*, IEEE Trans. SE–12, No. 9, 1986.

4.11   Test Adequacy Measures

Aim: To determine the level of testing applied using quantifiable measures.

Description: Testing proceeds until some pre-defined *stopping criterion* is reached. A range of stopping criteria can be used, such as:–

- Path coverage measures (e.g. 99% of all branches exercised).
- Detection and removal of a set of artificial faults seeded into the software (bug seeding).
- The proportion of mutated software versions that are detected with the test data suite.
- Number of tests without detected fault.
- Number of faults removed (as compared to a predicted level).

Conditions: The criterion should be related to the required quality of the software.

Major Advantages: A measurable target, achievement of the target can be demonstrated.

Problems or Disadvantages: There is no obvious relationship between the measure (e.g. branch coverage) and the desired product attributes (e.g. number of residual faults).

Related Methods: Related to all types of test activity.

Results: Documented test termination decisions.

Tools: Tools exist for structural coverage measures.

References:

1. *A Comparison of Structured Test Coverage Metrics*, M.D. Weiser, J.D. Gannon and P.R. McMullin, IEEE Software, Vol. 2, No. 2, Mar 1985.

2. *On the Statistical Validation of Computer Programs*, H.D. Mills, FSC-72-6015, IBM Federal Systems Division, Gathersburg, 1972.

3. *The Portable Mutation Testing Suite*, T.A. Budd, Univ. Arizona, Tech. Rep. TR83-8, Mar 1983.

4. *Assessing a Class of Software Tools*, M. Hennel, D. Hedley and I.J. Riddell, Proc. 7th Int. Conf. Software Engineering, Mar 1984.

5. *An Applicable Family of Data Flow Testing Criteria*, P.G. Frankl, E.J. Weyuker, IEEE Trans. SE-14, No, 10, Oct 1988.

4.12   Verification and Validation

Aim: Verification — to build the product right; validation — to build the right product.

Description: Used primarily to check that the software implements its specified logical function. An important part of software quality control. Verification makes use of a variety of techniques to ensure that:

- The items produced at each phase of the development are internally consistent.
- Each phase is consistent with the previous phase.

Validation uses a variety of techniques to ensure that the final product is fit for its intended environment.

Conditions: Must form part of an overall quality control plan. Personnel engaged in these activities should be independent from the software developers.

Major Advantages: Detects faults early, minimizes costs.

Problems or Disadvantages: The performance and resource costs of some techniques are not known. For high-integrity applications a similar amount of effort may be devoted to verification as development.

Results:

- A better product with fewer residual faults.
- Evidence of careful development to support the licensing process.
- Fault detection information to support reliability prediction.

Related Methods: Overlaps with software quality control and testing.

Assessment: Essential for safety-related systems.

Tools:

Verification:

- Program proof checkers.
- Program analysis tools.
- Structural complexity measures.
- Standards for verifiable software design and implementation.
- Standards conformance checking tools.
- Test harness and test generators.
- Test coverage measurement tools.

Validation:

- Prototyping tools.
- Plant simulators.

References:

1. *Techniques for the Verification and Validation of Safety-related Software.* in 'Dependability of Critical Computer Systems – Part 1', E.F. Redmill (ed.), Elsevier Applied Science, 1988.

2. *Verification and Validation of Real-Time Software.* W.J. Quirk (Ed.), Springer-Verlag, Feb. 1985.

3. *Formal Program Verification using Symbolic Execution.* R.B. Dannenberg and G.W. Ernst, IEEE Trans Software Engineering, Vol. SE-8, No. 1, 1982.

Chapter 5

Failure Detection

5.1   Control Flow Checks

Aim: To detect computer mal-operation by detecting deviations from the intended control flow.

Description: The structure of the program is analysed from the source or taken from the design documentation. Based on this analysis, additional code is inserted at key points in the program to check that the control flow sequence through the key points is valid. The choice of key points can vary, but would typically be:-

- subroutine entry and return points,
- branch points such as IFs and DO loops.

The control flow checking method can also vary, one method is to use a global trace variable which is updated with a unique number as it passes each key point. The key point code can check that the prior key point value is valid before performing its own update. If the control flow is found to be invalid, a failure containment strategy is invoked.

Major Advantages:

- Easy to implement.
- Detects some hardware mal-operations.
- Guards against some programming and compiler/linker faults.

Problems or Disadvantages:

- Additional memory space and timing overheads.
- The effectiveness of the technique is unquantified.

Results: A greater assurance that the system will be fail-safe.

Related Methods: A defensive programming technique, can be one of the integrity checks used to disable a watchdog timer.

Assessment: Not necessary if the basic hardware is fully proven and self-checking, and there is high confidence in the correctness of the program and the support tools.

If these conditions are not satisfied, it is a valuable technique for systems which can fail to a safe state where there is:-

- no hardware redundancy, or
- no software diversity in the program or support tools.

Tools: Program structure analysers may aid the designer in selecting suitable points for the insertion of run-time control flow checks:

- Logiscope^{TM} (Verilog, rue Nicolas Vauquelin, 31081 Toulouse, Cedex, France).
- Malpas^{TM} (Rex Thompson Partners, Newnhams, West Street, Farnham GU9 7EQ, England).
- Testbed^{TM} (Liverpool Data Research Associates, Liverpool Innovation Centre, 131 Mount Pleasant, Liverpool L3 5TS, England.

References: not available.

5.2 Data Security

Aim: To guard against external and internal threats which can either accidentally or deliberately endanger the objectives of design and may lead to unsafe operation.

Description: During the operational phase of a system, the potential threats and counter measures can be broken down into several areas:-

- communication
- files
- verification of data
- authentication of messages
- manipulation
- program load.

Communication and data exchange through interfaces requires protection due to the relatively poor reliability in these phases of information flow. Following approaches for coding are used in telecommunications for error detection and correction.

- parity check coding
- binary cyclic codes
- cyclic redundancy check codes
- BCH (Bose-Chandhuri-Hocquenghem) codes
- convolutional codes
- burst error codes
- coding against fading or interference.

The introduction of ISDN (integrated services digital network) forces the application of encryption and public keys even for public communication networks. These measures are needed for granting data security (to defend against unauthorized users), authentication of messages, manipulation detection, and for verifying data.

The security standards should address:–

- physical aspects
- programming, software and data
- personnel
- contingency planning
- internal control and organization.

Major Advantages: Reduces risk of inadvertent or deliberate damage to system data.

Problems or Disadvantages: Extra computing overheads, unless the system has hardware support.

Results: Increases system safety.

Assessment: Essential for safety-related systems.

Tools: Hardware and software encoders.

References:

1. *Data encryption algorithm*. ANSI X3.92-1981.

2. *Information processing, parity check against input errors*. ISO 7064 1982.

3. *Computers and security*. H.J. Highland (ed.), (Official IFIP TC11 Journal). Elsevier Science Publ., Amsterdam.

4. *Handbook for information security*. H.J. Highland (ed.), North-Holland, Amsterdam, 1985.

5. *The technology of error-correcting codes*. E.R. Berlekamp, Proc. of the IEEE, 68(1980) 5.

6. *Kodierung für die Datenubertragung und Fehler-sicherung*. K. Einer, Signal und Schiene, 20 (1976) 12.

7. *A design concept for reliable mobile radio networks with frequency hopping signalling*. A. Epheremides, J.E. Wieselthier and D.J. Baker, Proc. of the IEEE, 75 (1987) 1.

8. *A short course on error correcting codes*. N.J.A. Sloane, Springer-Verlag, Wien, 1975.

9. *Error-detecting codes, self-checking circuits and applications*. J. Wakerly, North-Holland, Amsterdam, 1978.

5.3   Dynamic Logic

Aim: To provide self-supervision by the use of a continuously changing signal.

Description: For hardware systems, additional pulses are introduced into the system, causing switching of the logic elements. A static state is considered to be a hardware failure.

For software, additional counters are introduced into the system, which should increment within a certain period. The state of the counters has to be continuously checked. The dynamic software state is usually linked to some hardware dynamic logic signal.

Conditions: For software this method requires knowledge of the maximum interval between changes to the counters. For hardware, this requires some form of pulse generator and dynamic logic decoding circuitry on the output. The decoder should also be fail-safe i.e. fail to a static rather than a dynamic state.

Major Advantages: Early detection of hardware failure; limited level of software failure detection; may in some cases permit a non-redundant system to be used if the plant has a fail-safe state.

Problems or Disadvantages: Some extra equipment needed. Need to demonstrate failure modes are indeed static. Some extra memory and computing overhead for the software.

Limitations: No protection against most software design faults.

Related Methods: Fail-safe design; plausibility checks.

Assessment: Essential in non-redundant systems. Desirable in redundant systems as a means of distinguishing faulty channels.

Tools: None.

References:

1. *Systeme de Logique Dynamique pour les Circuits de Protection des Réacteurs Nucléaires.* S. Salomon and J.C. Therond, NUCEX 75, Merlin Gerin, Oct. 1977.

2. *Transition Count testing of Combinational Logic Circuits.* J.P. Hayes, IEEE Transactions on Computer., Vol. c-25, no. 6, June 1976.

3. *A fail-safe computer-based reactor protection system.* J. Wakerly, IAEA Specialists meeting on use of computers for protection systems and automatic control, Munich, May 1976.

4. *Design for Testability- A survey.* T. W. Williams, and K.P. Parker, Proc. of IEEE Vol. 71, No. 1, Jan. 1983.

5.4   Invariant Assertions

Aim: To detect whether a computer system has deviated from its intended function.

Description: The objective is to make an assertion about either the software or the observed state of the plant which is *always* true under normal operational conditions. In software the assertions could be derived directly from a formal software development process, where the pre-conditions and post-assertions for each module are fully defined. However it is also possible to define more general assertions. For example, assertions could relate only to pre-specified safety boundaries, so that it may not detect incorrect operations, only unsafe ones. Similarly the assertion can be a simple plausibility check on the results of the software or the state of the system hardware or the process information. The invariant assertions are checked at run-time and if a failure is detected, some failure containment strategy is invoked.

Conditions: No critical time-response requirements. Good methods of defining invariant assertions (such as formal specification methods).

Major advantages: Increases confidence that when the system is running, it is running correctly (or at least safely). Can detect faults in the software, the software development tools and the underlying computing hardware.

Problems or disadvantages: Extra effort may be required to define and validate these assertions. This is very dependent on the degree of formality used during system definition development. The proportion of failures likely to be detected by this technique is not known.

Results: A system that is more likely to detect a failure condition.

Related Methods: Related to formal specification methods and fault containment techniques.

Assessment: To be used on non-time critical safety related systems.

Tools: None.

References:

1. *Assigning meaning to programs.* R.W. Floyd, Proc. Symp. Applied Math. Vol. 19.

2. *Math. Aspects of Computer Science.* R.W. Floyd, J.T. Schwartz et. al., American Math. Society, 19–32, 1967.

3. *The axiomatic basis of computer programing.* C.A.R. Hoare, ACM Trans. Vol. 12, Oct. 1969.

4. *A Discipline of Programming.* E.W. Dijkstra, Prentice Hall, 1976.

5.5   Memorizing Executed Cases

Aim: To force the software to fail-safe if it executes an unlicensed path.

Description: During licensing a record is made of all relevant details of each program execution. During normal operation each program execution is compared with the set of the licensed executions. If it differs, a safety action is taken.

The execution record can be the sequence of the individual decision-to-decision paths (DD paths) or the sequence of the individual accesses to arrays, records or volumes, or both.

Different methods of storing execution paths are possible. 'Nash'-coding methods can be used to map the execution sequence onto a single large number or sequence of numbers. During normal operation the execution path value must be checked against the stored cases before any output operation occurs.

Since the possible combinations of decision-to-decision paths during one program is very large, it may not be feasible to treat programs as a whole. In this case, the technique can be applied at the module level.

Conditions: Proper modularization of the supervised code. Extra storage is needed to hold stored executions (at least 3 times the storage requirement of the code). Ample additional execution time must be available to record the execution and check it against the stored cases. This can be more than double the execution time. There must be a safe state for the plant.

Major Advantages: The necessary instrumentation of the code for licensing and normal operation can be done automatically. For systems that work in co-operation with a human operator, the operator may add new cases among the memorized ones, after he has personally checked their correct treatment through the software.

Problems or Disadvantages: More memory space and computing time is required. If there are a large number of paths through the complete program, its application may have to be restricted to individual subroutines.

Related Methods: Related to testing and fail-safe design.

Assessment: There is little performance data available, so it is too early to make positive recommendations.

Tools: Not available yet.

References:

1. *Fail-safe Software — Some Principles and a Case Study.* W. Ehrenberger, in Proc. SARSS '87, Altrincham, Manchester, UK, 11–12 Nov. 1987, B.K. Daniels (ed.), Elsevier Applied Science, 1987.

5.6   Redundancy for Fault Detection

Aim: By employing redundancy, checks may be made for differences between units to determine sub-system failures.

Description: Redundancy may be employed to enable differences and hence failures to be determined. Dual redundant systems may compare their output results and a difference is taken as indicating that one of the two sub-systems has failed. However it is not possible to diagnose which unit has failed without further external checks. For higher levels of redundancy, majority voting (N out of M) techniques can be used to determine which unique sub-system item has failed.

Communication systems using redundancy may alternately send messages through different routes to determine that each communication path is fault free. Health states of a communication path may be established through complete communication failure or the number of re-tries due to errors that are required.

Conditions: Can be applied only if redundant units exist. A reliable discrepancy checking system is necessary.

Major Advantages:

- Rapid failure detection during normal operation.
- The fault coverage increases with time as the system is exposed to different operating conditions.
- When N out of M voting is employed, the faulty component can be diagnosed.

Problems or Disadvantages: Additional cost of redundancy and time for checks. Does not defend against common design faults.

Results: Fast and reliable failure detection.

Related Methods: Error detecting codes, self testing and capability checking, software time-outs.

Assessment: Should always be used in safety computer applications.

References:

1. *SIFT : System design and Implementation.* C. Weinstock, 10th Symposium on Fault Tolerant Computing, 1980.

2. *The Evolution of Fault-Tolerant Computing.* (Vol. 1 of *Dependable Computing and Fault Tolerant Systems*), Edited by A. Avižienis, H. Kopetz, and J.C. Laprie, Springer-Verlag, ISBN 3-211-81941-X, 1987.

3. *Fault Tolerance, Principles and Practice.* (Vol. 3 of *Dependable Computing and Fault Tolerant Systems*), T. Anderson and P.A. Lee, Springer Verlag, ISBN 3-211-82077-9, 1988.

4. *Fault Tolerant Computing Theory and Techniques.* Vol. 1 and 2, D.C. Pradhan (Ed.), Prentice-Hall, 1986.

5. *Reliable and Safe Control System Using Fault Tolerance.* J. Wensley and C.J. Goring, 9th Advancements in Reliability and Safety Symposium, Bradford, UK, 1986.

6. *Availability by Replication in Embedded Distributed Systems.* A. Corradi, L. Leonardi and C. Salati., in Proc. SARSS '87, Altrincham, Manchester, England, 11–12 Nov. 1987, B.K. Daniels (ed.), Elsevier Applied Science, 1987.

5.7   Self Testing and Capability Testing

Aim: To verify on-line that the system maintains its capability to act in the correct and specified manner.

Description: By forcing input states in an ordered and known sequence whilst monitoring the output reaction (final output switching element disabled or reaction pulsed) the complete functionality of the system is checked. Output data is validated to the expected known result. Diagnostic and repair action is initiated on failure.

Conditions: Design techniques need to be employed to enable outputs to be fully checked. Test programs must be practical in both time and added complexity. All tests may be aborted (over-ridden) if any real demand input state change occurs.

Major Advantages: Regular assurance of system functionality and hence increased system availability.

Problems or Disadvantages:

- Increased hardware design and complexity.
- Increased software design complexity.
- Only effective against permanent rather than transient faults.
- With a fixed self-test scheme, the fault coverage is fixed and the actual coverage obtained is difficult to measure.

Results: Reduces the potential for latent or hidden faults and hence reduces the incidence of hazard situations.

Related Methods: Standard error detecting techniques employed as part of self test; these include voting, parity, error checking and correcting code etc.

Assessment: Essential on a normally dormant primary safety system.

Tools: None.

References:

1. *Latent Fault Detection in Fault Tolerant Computer Based Safety and Control Systems.* C.J. Goring, Proc. SARSS '87, Altrincham, Manchester, UK, 11–12 Nov. 1987, B.K. Daniels (ed.), Elsevier Applied Science, 1987.

2. *On Line Diagnostics for an Industrial Microcomputer Eases the Maintenance Burden.* J.P.D. Wales, 1982. Symposium on Microcomputer Applications in Power Engineering, IEE – North East – Univ. Newcastle, UK.

3. *Error-detecting codes, self-checking circuits and applications.* J. Wakerly, North-Holland, Amsterdam, 1978.

4. *A Unified Built-in Self Test Scheme: UBIST.* M. Nicolaidis, FTCS 18, Tokyo, 27–30 July 1988, Tokyo, IEEE, 1988.

5.8   Software Time-Out Checks

Aim: To provide time limits for software running non-deterministic tasks.

Description: Many internal and external communication paths, be it inter-task communication within a processor subset or inter-application program communication from one processor to another, cannot be defined in a deterministic manner. Software time-outs provide a method of limiting the maximum time allowed for these tasks, allowing decisions such as abort or re-try to be taken on the occurrence of a time-out.

Conditions: An implicit knowledge of the permissible times that the relevant tasks will take is required.

Major Advantages: May be used to provide levels of fault tolerance, an example would be to re-try and re-configure a network route on a communication time-out.

Disadvantages: Minor software overhead needed.

Results: The ability to control and provide determinism on non-deterministic task.

Assessment: Should always be used to provide determinism on non-deterministic task in safety computer systems.

Related Methods: Related to error recovery techniques and watchdog timers.

Tools: None known.

References:

1. *Timing constraints for real-time systems: constructs for expressing them, methods of validating them.* T.L. Booth et. al., Proc. IEEE real-time systems symp., Dec. 1982.

2. *Introducing real-time constraints into requirements and high level design of operating systems*. B.J. Taylor, IEEE conf. rec., National telecomms. conf., Dec. 1980.

3. *An operational approach to requirements specification for embedded systems*. P. Zave, IEEE Trans. Soft. Eng., Vol. SE-8, No. 5, May 1982.

5.9 Use of Assertions and Plausibility Checks

Aim: Produce code whose intermediate results are continuously checked during execution. In case of incorrect results a safety measure is taken.

Description: Intermediate results are checked for correctness, validity or plausibility during execution.

For example:-

>   ASSERT *precondition*;
>   action 1;
>   action 2;
>   ....
>   action n;
>   ASSERT *postcondition*;

Conditions: Any problem whose loss of information content during execution is high, qualifies for this method. Careful analysis of the predicates at the different distinct steps of program execution is necessary. Actions have to be planned to deal with a negative outcome of an assertion.

Major Advantages: Checks during on line execution. Recovery is possible in some cases. May be useful as a means of detecting hardware faults and compiler faults as well as faults in the software.

Problems or Disadvantages: No guarantee of correctness can be given. Sometimes only very vague predicates that do not mean very much can be given. Derivation of predicates requires full understanding of problem and program; cannot be automated at present. Only reasonable if a safe state exists. Predicates are less useful if the program is correct or if it has been verified by other means.

Related Methods: Related to self testing and capability checking.

Assessment: Recommended, if no complete test or analysis is feasible; conditions mentioned apply; problem is fairly complex.

Tools: Some artificial intelligence languages provide means for easy formulation of the assertions.

References:

1. *A discipline of programming.* E.W. Dijkstra, Prentice-Hall, 1976.

2. *The science of programming.* D. Gries, Springer-Verlag, 1981.

3. *Software development — a rigorous approach.* C.B. Jones, Prentice-Hall, 1980.

5.10   Watchdog Timers

Aim: To provide a non-software related reliable hardware checking method of the software operation.

Description: Watchdog timers provide hardware timing circuits configured as simple re-triggerable mono-stables timers, window (dual inverted mono-stables) timers and cascaded window timers. Within the application software, on the completion of critical task or scans, an output to the relevant watchdog circuit re-sets the hardware timers. Should a critical task fail to re-set the related watchdog within its configured timespan, the hardware circuit times-out and provides a physical alarm signal (e.g. relay contact) which may be hard wired to perform corrective order functions.

Conditions: Watchdog timer circuits should be physically unrelated to the processor timing circuits. The only interaction should be the relevant output driver to re-set the watchdog timer. An understanding of system task timing is a prerequisite.

Major Advantages: Properly configured, watchdog circuits can provide confidence that critical tasks are being run and a hardware indication that no major computer system malfunction has occurred.

Problems or Disadvantages: There is a minor software overhead in servicing the watch-dog task and of course a minor hardware overhead in supplying the circuit.

Results: Task implementation confidence is provided. Processor health confidence is provided.

Related Methods: Related to software time-out checks but provide an extra 'hardware' level of time-out security, related to error recovery techniques.

Assessment: Should be used on all safety critical and real-time control systems.

Tools: None known.

References:

1. *A Top Down Approach to Safety Systems.* (Industrial Control Services) Control and Instrumentation, November 1986.

2. IFIP-ICC Vocabulary of Information Processing, 1966.

3. *Microprocessors in Instruments and Control.* Robert J. Bibbero, Wiley Interscience, 1977.

4. *Microprocessor Fundamentals : Hardware and Software.* Ramirez, McGraw-Hill Book Company GmbH, 1986.

5. *An Introduction to Microcomputers, some real microprocessors.* Vol I and II, J. Kane, A. Osborne, Osborne Associates, Inc., 1978.

Chapter 6

Failure Containment

6.1   Defensive Programming

Aim: To produce programs which detect anomalous control flow, data flow or data values during their execution and react to these in a predetermined and acceptable manner.

Description: Many techniques can be used during programming to check for control or data anomalies. These can be applied systematically throughout the programming of a system to decrease the likelihood of erroneous data processing.

Two overlapping areas of defensive techniques can be identified. Intrinsic error-safe software is designed to accommodate its own design shortcomings. These shortcomings may be due to plain errors of design or coding, or to erroneous requirements. The following lists some of the defensive techniques.

- Variables should be range checked.

- Where possible, correlated variables should be checked for plausibility.

- Parameters to procedures should be type, dimension and range checked at procedure entry.

  These first three recommendations help to ensure that the numbers manipulated by the program are reasonable, both in terms of the program function (e.g. only take square root of positive numbers) and physical significance of the variables.

- Read-only and read-write parameters should be separated and their access checked. Functions should treat all parameters as read-only. Literal constants should not be write accessible. This helps detect accidental overwriting or mistaken use of variables.

Error tolerant software is designed in as far as is practicable to 'expect' failures in its own environment (e.g. anomalous inputs) or use outside nominal or expected conditions, and behave in a predefined manner. Techniques include the following.

- Input variables and intermediate variables with physical significance should be checked for plausibility.

- The effect of output variables should be checked, preferably by direct observation of associated system state changes.

- The software should check its configuration. This could include both the existence and accessibility of expected hardware (which could be considered in section 3 below) and also that the software itself is complete (e.g. all ROMS present, of the same release/version and in the correct order — N.B. a simple checksum is not sufficient). This is particularly important for maintaining integrity after maintenance procedures.

Some of the defensive programming techniques, such as control flow sequence checking, also cope with external failures.

Major Advantages:

- Coverage of hardware and software failures.
- Fairly easy to use and relatively low cost.

Problems or Disadvantages: It is difficult to quantify or assess the defensive coverage gained by using the technique and it should not be relied upon solely. It may also increase significantly the size of the code, and may slow down the execution speed.

Results: Increased confidence in the function performed by the program.

Related Methods:

Defensive methods can use a variety of failure detection methods and may be used in conjunction with any failure handling strategy. With a fail-stop approach, detected anomalies can be designed to de-activate a watchdog timer.

Assessment: Recommended where there is insufficient confidence in the environment or the software. Plant sensor and actuator checks are probably essential. Proven fail-stop hardware, support tools and application software, or redundancy coupled with software diversity may make some of the other checks unnecessary.

Tools:

High level languages such as ADA provide exception handling mechanism which can simplify the implementation of defensive programming.

References:

1. *Dependability of Critical Computer Systems – Part 1.* E.F. Redmill (ed.), Elsevier Applied Science, 1988.

2. *Dependability of Critical Computer Systems – Part 2.* E.F. Redmill (ed.), Elsevier Applied Science, 1989.

3. *Software Engineering Aspects of Real-time Programming Concepts.* E. Schoitsch, Computer Physics Communications, 41 (1986), North Holland, Amsterdam.

6.2   Diversity

Aim: To enhance the safety of a system by applying separate systems that simultaneously perform safety critical tasks.

Description: Diversity is used to prevent a single functional error or a design error or other systematic error having a safety threatening effect. The diversity is applicable at many different levels:

- functional (e.g. to have trip activation on both pressure and temperature limit),

- design (e.g. to let independent teams design the same device in parallel channels), and

- software (implementing N versions to the same specification)

- manufacturing (e.g. by using chips from different manufacturers in parallel channels).

Conditions: Sufficient resources must be available.

Major Advantages: Reduces the probability of common cause failures, namely those failures which are caused by the same inherent fault in each parallel channel of a redundant system.

Problems or Disadvantages:

- Increased cost.
- It may be difficult to compare the results from diverse devices.
- The voting mechanism between the results from the diverse devices may in itself be vulnerable to common cause failures.

- Demonstration of the independency of the diverse solution may be difficult.

Related Methods: Similar to, but more powerful than, redundancy without diversity. Complementary to geographical separation.

Assessment: Recommended for safety-critical systems.

Tools: None known.

References:

1. *A Defense in Depth and Diversity Assessment of the RESAR-414 Integrated Protection System.* NUREG, Division of System Safety Office of Nuclear Reactor Regulation U.S. Nuclear Regulatory Commission Washington, D.C. 20555, March 1979.

2. *Exigence de Sureté relative au logiciel utilise dans les systemes de securite des reacteurs nucleaires.* J.M. Collart and J. Grisollet, Commissarat a L'Energie Atomique, Institut De Protection et de Sureté Nucleaire, Department de Sureté Nucleaire, Service D'Analyse Fonctionnelle, Rapport S.A.F. N0 32, 28 July 1982.

3. *Reactor Protection System Diversity in Westinghouse Pressurized Water Reactors.* T.W. Burnett, Westinghouse Electric Corporation, Topical Report WCAP-7306, April 1969.

4. *Criteria for Protection Systems for Nuclear Power Generating Stations.* IEEE Std 279-1971, The Institute of Electrical and Electronic Engineers, IEEE 345 East 47th Street New York N.Y., USA.

6.2.1   Diversity: N-version Programming

Aim: Detect and mask residual software design faults during execution of a program, in order to prevent safety critical failures of the system, and to continue operation for high reliability.

Description:

In N-version programming a given program specification is implemented N times in different ways. The same input values are given to the N versions, and the results produced by the N versions are compared. If the result is considered to be valid, the result is transmitted to the computer outputs.

The N versions can run in parallel on separate computers, alternatively all versions can be run on the same computer and the results can be subjected to an internal vote. Different voting strategies can be used on the N versions, depending on the application requirements:-

- If the system has a safe state, then it is feasible to demand *complete* agreement (all N agree) otherwise a fail-safe output value is used. For simple trip systems the vote can be biased in the safe direction (i.e. trip). In this case the safe action would be to trip if either version demanded a trip. This approach typically uses only two versions ($N = 2$).

- For systems with no safe state, majority voting strategies can be employed. For cases where there is no collective agreement, probabilistic approaches can be used (e.g. taking the middle value, temporary 'freezing' of outputs until agreement returns, etc) in order to maximise the chance of selecting the correct value.

This method does not eliminate residual software design faults, but it provides a measure to detect and mask them before they can affect safety.

Conditions: Sufficient resources must be available.

Major Advantages:

- The probability of common cause failures is reduced. Residual design and coding faults can be detected and masked at run-time without loss of service.
- The availability of multiple versions permits use of comparison testing with a large number of test cases as part of the software acceptance (and re-acceptance procedures).

Problems or Disadvantages:

- If only a single specification is used, specification errors may remain undetected.
- Comparison of N versions may be difficult.
- Additional software for voting and synchronization is necessary; voting and synchronization in itself may cause common mode failures.
- Increased development and maintenance cost compared to a single version implementation.

Related Methods:

See other models of design diversity (recovery block scheme, assertion programming, safety bag technique). Complementary to physical (geographical) separation and redundancy by identical replication.

Assessment:

Recommended for safety relevant fault compensating systems. At least one form of design diversity should be applied for safety critical systems possibly affecting human life.

Tools: Not applicable.

References:

1. *Dependable Computing: From Concepts to Design Diversity.* A. Avižienis, J.C. Laprie, Proc IEEE, Vol. 74, 5, May 1986.

2. *A Theoretical Basis for the Analysis of Multi-version Software subject to Co-incident Failures.* D.E. Eckhardt and L.D. Lee, IEEE Trans. SE-11, no. 12, 1985.

3. *An Experimental Evaluation of the Assumption of Independence in Multi-version Programming.* J.C. Knight and N.G. Leveson, IEEE Trans. SE-12, no. 1, 1986.

4. *Software Diversity in Computerized Control Systems.* (Vol. 2 of *Dependable Computing and Fault Tolerant Systems.* U. Voges (Ed.), Springer-Verlag, ISBN 3-211-82014-0, 1987.

5. *PODS – A Project on Diverse Software.* P.G. Bishop et al, IEEE Trans., SE-12, no. 9, 1986.

6. *PODS Revisited – A Study of Software Failure Characteristics.* P.G. Bishop and F.D. Pullen, FTCS 18, Tokyo, 27–30 July 1988, Tokyo, IEEE, 1988.

7. *Error Masking: A Source of Dependency in Multi-Version Programs.* P.G. Bishop and F.D. Pullen, IFIP 10.4 International Working Conference on "Dependable Computing for Critical Applications", University of California, Santa Barabara, August 23–25, 1989.

8. *In Search of Effective Diversity: A Six Language Study of Fault-tolerant Flight Control Software.* A. Avižienis, M.R. Lyu and W. Schültz, FTCS 18, Tokyo, 27–30 July 1988, Tokyo, IEEE, 1988.

6.2.2   Diversity: The Safety Bag

Aim:

To protect against residual specification and implementation faults in software which adversely affect safety.

Description:

A 'safety bag' is an external monitor, implemented on an independent computer to a different specification. This safety-bag is solely concerned with ensuring the main computer performs safe (not necessarily correct) actions. The safety bag continuously monitors the main computer. The safety bag prevents the system from entering an unsafe state. In addition if it detects that the main computer is entering a potentially hazardous state, the system has to be brought back to a safe state either by the safety bag or the main computer.

Conditions: Only applicable to fail-safe systems.

Major Advantages:

- The specifications for the main computer and the safety bag differ, hence reducing the chance of common specification-related faults.
- Can be used on real-time systems.
- The safety bag software is normally simpler than the software in the main computer, hence reducing the cost of development compared with 2-version diversity.

Major Problems or Disadvantages:

- There is no fault masking, so both computers have to be operational to maintain availability.
- Increased maintenance and development costs compared with a single version.

- There may be problems with information exchange and voting.
- There is no protection against failures in the safety bag.

Related Methods: A form of diversity. Related to assertion programming. Similar in some ways to the recovery block scheme (without the recovery).

Assessment: A relatively novel approach. Should be considered for fail-systems, provided there is adequate confidence in the dependability of the safety bag itself.

Tools: Not applicable.

References:

1. *Using AI Techniques to Improve Software Safety.*
   N. Theuretzbacher, Proc. IFAC SAFECOMP 86, Sarlat, France, Oct. 1986, Pergamon Press 1986.

6.3   Error Correcting Codes

Aim: To detect and correct errors in sensitive information.

Description: For an information of n bits, a coded block of k bits is generated which enables r errors to be detected and corrected. Different types of code include:

- Hamming codes
- cyclic codes
- polynomial codes

Conditions: Encoding and decoding facilities (implemented in either hardware or software).

Major Advantages: In transmission systems, error correction is immediate, avoiding delays associated with re-transmission. In memory systems, random transient and memory chip failures can be tolerated.

Problems or Disadvantages: Additional time is required for message transmission or for memory error checking and correcting. Check blocks may become very large. Check codes designed to detect errors rather than correct errors are more efficient. The error correcting code technique is usually applied to data transmission systems and computer memory storage systems where time delays are critical. In principle it could also be implemented in software to protect sensitive blocks of data, but it should only be applied to individual words rather than checking the consistency of the whole data block, again because the check blocks become prohibitively large.

Results: Reduction in the number of data errors, correction delays are minimized.

Related Methods: Information encoding, error detecting codes.

Assessment: Not as generally applicable as normal error detecting codes, but may be useful in systems where availability and response times are critical factors.

Tools: Encoders, decoders.

References:

1. *The technology of error-correcting codes*. E.R. Berlekamp, Proc. of the IEEE, 68, 5, 1980.

2. *A short course on error correcting codes*. N.J.A. Sloane, Springer-Verlag, Wien, 1975.

6.4 Fail Safety

Aim: To design a system such that failures will drive the system to a safe state.

Description: This is a general design principle which can be applied to both hardware and software. The safe states (if any) of the plant are identified, together with the potential failure modes of the control system. The system is then designed such that known failures in the automation system will drive the plant to a safe state to very high probability. In the ideal system no form of failure detection is required to implement a fail-safe design, since the bias towards safety is an inherent part of the design. A typical example of fail-safe hardware design is a vehicle braking system where the power is used to hold the brakes <u>off</u>. Fail-safe software designs tend to rely on the continuous generation of some form of 'health signal' which is unlikely to be generated by accident. These health signals may rely on some external hardware detection circuit which must itself be fail-safe. Typical examples are systems where a software-generated pulse is checked by a hardware watchdog, or systems where only specific software-generated dynamic patterns are regarded as valid by the external input/output logic.

The principle is also applicable to fault tolerant systems using active failure detection features. A safe action may be used if there is no common agreement, or when there is a complete loss of function.

Conditions: The plant must have safe states. The failure modes of the automation system must be known.

Major advantages:

- Can assure safety to a high probability without active intervention or failure detection.

- Can be linked to other forms of internal self test.

Problems or disadvantages:

- The plant may not have safe states, or safe states may not be reachable in all circumstances.
- A fail-safe design may result in unacceptably low levels of availability; this can be mitigated by fault tolerance techniques such as redundancy and majority voting.
- Fail-safe designs require considerable design and analysis effort.

Results: An automation system with a strong bias towards safe rather than unsafe modes of failure.

Related Methods: Methods such as watchdogs, dynamic logic, vital coded processing and failure tracking can be used to implement fail-safe designs. FMEA and FMECA can be used in the design process. Similar techniques may be used to implement 'fail-silent' components of a fault tolerant system architecture.

Assessment: Strongly recommended for systems where there are safe plant states.

Tools: Hardware watchdogs, pattern recognition circuits.

References:

1. *Programmable Electronic Systems in Safety Related Applications: "1 An Introductory Guide"*, Health and Safety Executive, 1987 ISBN 011 8839136, HMSO, PO Box 276, London SW8 5DT.

2. *Programmable Electronic Systems in Safety Related Applications: "2 General Technical Guidelines"*, Health and Safety Executive, 1987, ISBN 011 8839063, HMSO, PO Box 276, London SW8 5DT.

3. *Fail-safe Control Systems*, Unicom Seminar, 28-30 June 1988, Brunel Science Park, Cleveland Road, Uxbridge, England.

6.5   Failure Tracking

Aim: To minimize the consequences of detected failures in the hardware and software.

Description: An error flag is associated with all data items held in the computer. When a new value for a data item is computed, the error flag is set if the computation fails either because:-

- the computation uses another data item which has an error flag set.
- computational overflow or underflow is detected.

The concept can also be applied to shared data used in a multi-tasking and distributed processor environment. In this case the error flag may be set if the process or processor supplying the information is known to have failed (e.g. detected by time-out).

The error flags are allowed to propagate through the system, and may be inspected by a process in order to take corrective action. If no action is taken, attempts to control system interfaces with error-flagged data will result in some nominated default action (usually 'freeze' or disable) which should be fail-safe.

Conditions:

- Requires a special language translator or pre-processor in order to implement failure tracking automatically.
- Requires good failure detection techniques on system input/output, process computation, communication systems and processors.

Major Advantages: Should result in a more fail-safe design. Reduces the probability that unexpected failures will cause a hazard.

Problems and Disadvantages: Software is slower and needs more memory than systems without failure tracking. Dependent on the quality of the fault detection methods.

Results: Reduced hazard to the system.

Related Methods: One option for the management of failure, may use a range of fault detection techniques as the basis for setting the error flags.

Assessment: Desirable for safety-related applications, but alternative approaches to failure management may also be used.

Tools: CUTLASS (a language to support process control applications).

References:

1. *Fault Detection and Recovery Techniques in CUTLASS.* P.G. Bishop, IFAC Workshop, SAFECOMP 83, Cambridge, UK, September 1983, Pergamon Press (Oxford), 1983.

2. *CUTLASS Overview Manual.* 1988, Central Electricity Generating Board (CISD), 15 Newgate St., London EC1, UK.

6.6   Fault Tolerance

Aim: To provide correct functional operation (or at least the essential part of it) in the presence of one or more faults.

Description: The configuration of system components is designed using both redundancy and diversity, to mask one or more faults from the system operation. The level of fault tolerance required will determine the level and sophistication of techniques that are applied, ranging from simple reset and re-try through dual redundancy to N out of M voting.

Conditions: To fully meet the 'aim' of fault tolerance the redundant paths must be isolated from each other with no hardware common mode failure paths. Similarly the interactions through communication paths should not be able to corrupt software in adjacent components.

On-line testability of all items to reveal latent faults must form an integral part of the design, as should the ability to repair and maintain faulty components without affecting the overall system functionality and availability.

Major Advantages: Correctly applied fault tolerance techniques provide a significant improvement in system availability and hence in the potential for safer operation of the system.

Problems or Disadvantages: All aspects of system design with respect to common mode failures, initial failures and repairability must be carefully considered and therefore additional complexity of system design and integration inevitably occurs.

Results: A very high availability from the control and safety systems can be obtained.

Related Methods: Fault Tolerance should always be

considered in relationship to the total system design encompassing actuators, sensors and transmitters. Will make use of some form of redundancy (and possible diversity) and voting strategies, including dynamic reconfiguration and graceful degradation.

Assessment: Essential for any system without a safe state or for systems which have a high cost associated with loss of availability.

Tools: Fault Tree analysis tools, Markov modelling tools. Proprietary fault-tolerant hardware (Tandem, Stratus, Bonar-August etc).

References:

1. *The Evolution of Fault-Tolerant Computing.* (Vol. 1 of *Dependable Computing and Fault Tolerant Systems*), Edited by A. Avižienis, H. Kopetz, and J.C. Laprie, Springer-Verlag, ISBN 3-211-81941-X, 1987.

2. *Fault Tolerance, Principles and Practice.* (Vol. 3 of *Dependable Computing and Fault Tolerant Systems*), T. Anderson and P.A. Lee, Springer Verlag, ISBN 3-211-82077-9, 1988.

3. *Fault Tolerant Computing Theory and Techniques.* Vol. 1 and 2, D.C. Pradhan (Ed.), Prentice-Hall, 1986.

4. *Reliability Prediction of Electronic Equipment.* MIL-HDBK-217.

6.6.1   Dynamic Reconfiguration

Aim: To maintain system functionality despite an internal fault.

Description: The logical architecture of the system has to be such that it can be mapped onto a subset of the available resources of the system. The architecture needs to be capable of detecting a failure in a physical resource and then remapping the logical architecture back onto the restricted resources left functioning. Although the concept is more traditionally restricted to recovery from failed hardware units, it is also applicable to failed software units if there is sufficient 'run-time redundancy' to allow a software re-try or if there is sufficient redundant data to make the individual and isolated failure of little import.

Although traditionally applied to hardware, this technique is being developed for application to software and, thus, the total system. It must be considered at the first system design stage.

Conditions: To be applicable, the concept must be included from the inception of the system. The specification and design must accommodate it, and redundant resources must be included.

Major Advantages: Much increased system availability at full functionality if the potential is fully exploited.

Problems or Disadvantages: Achieving the full potential is difficult and has proved to be error prone itself. The correctness of error recovery software is acknowledged as being hard to establish. There is also an increased system cost for both extra resources and extra software.

Results: System availability increased substantially.

Related Methods: This is the strongest variety of reconfiguration, being essentially an automated enhancement to 'hot standby', 'warm re-start' and other similar techniques. It is also stronger than 'graceful degradation', since full functionality is maintained.

Assessment: Valuable where high fault tolerance and high availability are both required, but costly and difficult to validate.

Tools: None at present.

References:

1. *Critical Issues in the Design of a Reconfigurable Control Computer*. H. Schmid, J. Lam, R. Naro and K. Weir, FTCS 14 June 1984, IEEE 1984.

2. *Assigning Processes to Processors: A Fault-tolerant Approach*. June 1984, G. Kar and C.N. Nikolaou, IBM T.J. Watson Research Center, Yorktown Heights N.Y. 10598, J. Reif, Aikei Computation Laboratory, Harvard University, Cambridge, Massachusetts 02138, USA.

3. *Reconfiguration Procedure in a Distributed Multiprocessor system*. G. Barigazzi, A. Ciuffoletti and L. Strigini, Proceedings of FTCS-12, IEEE, 1982.

4. *Dynamically Reconfigurable Fault Tolerant Array Processors*. J.A.B. Fortes and C.S. Raghavendra, Department of Electrical Engineering-Systems, University of Southern California, Los Angeles, CA 90089, USA.

5. *The Design, Analysis and Verification of the SIFT Fault Tolerance System*. J.H. Wensley, 1st Conf. on Software Eng. San. Francisco Oct. 1976.

6. *The Mecra: A Self Reconfigurable Computer for Highly Reliable Process*. F.P. Maison, IEE Trans. Vol. C-20, No. 11, Nov. 1971.

7. *The Evolution of Fault-Tolerant Computing.* (Vol. 1 of *Dependable Computing and Fault Tolerant Systems*), Edited by A. Avižienis, H. Kopetz, and J.C. Laprie, Springer-Verlag, ISBN 3-211-81941-X, 1987.

8. *Fault Tolerance, Principles and Practice.* (Vol. 3 of *Dependable Computing and Fault Tolerant Systems*), T. Anderson and P.A. Lee, Springer Verlag, ISBN 3-211-82077-9, 1988.

9. *Fault Tolerant Computing Theory and Techniques.* Vol. 1 and 2, D.C. Pradhan (Ed.), Prentice-Hall, 1986.

10. *Reconfiguration in Microprocessor Schemes.* F. Lombardi, Microprocessing and Microprogramming, Vol. 13, 1984.

6.6.2 Graceful Degradation

Aim: To maintain the more critical system functions available despite failures by dropping the less critical functions.

Description: This technique gives priorities to the various functions to be carried out by the system. The design then ensures that should there be insufficient resources to carry out all the system functions, then the higher priority functions are carried out in preference to the lower ones. For example, error and event logging functions may be lower priority than system control functions. System control would continue if the hardware associated with error logging were to fail. Further, should the system control hardware fail, but not the error logging hardware, the error logging hardware would take over the control function.

This is predominantly applied to hardware but is applicable to the total system. It must be taken into account from the top-most design phase.

Conditions: The functionality of the system must be such that some parts of this may be withdrawn without the system performance becoming unacceptable.

Major Advantages: This approach usually costs less than full reconfiguration, particularly in its demands for physical resources.

Disadvantages: Assigning priorities to the various system functions may not be clear cut or easy to agree. The loss of some functionality nearly always has some impact, albeit not particularly serious. The consequences of a partial loss may only become apparent when it occurs.

Results: Critical function availability is increased at relatively low cost.

Related Methods: A cheaper reconfiguration strategy

than full dynamic reconfiguration, hot standby, etc.

Tools: None known.

Assessment: Highly recommended for systems with no fail-safe state.

References:

1. *Space Shuttle Software.* C.T. Sheridan, Datamation, Vol. 24, July 1978.

2. *The Evolution of Fault-Tolerant Computing.* (Vol. 1 of *Dependable Computing and Fault Tolerant Systems*), Edited by A. Avižienis, H. Kopetz, and J.C. Laprie, Springer-Verlag, ISBN 3-211-81941-X, 1987.

3. *Fault Tolerance, Principles and Practice.* (Vol. 3 of *Dependable Computing and Fault Tolerant Systems*), T. Anderson and P.A. Lee, Springer Verlag, ISBN 3-211-82077-9, 1988.

4. *Fault Tolerant Computing Theory and Techniques.* Vol. 1 and 2, D.C. Pradhan (Ed.), Prentice-Hall, 1986.

6.7   Recovery Blocks

Aim: To increase the likelihood of the program performing its intended function.

Description: Several different program sections are written, often independently, each of which is intended to perform the same desired function. The final program is constructed from these sections. The first section, called the primary, is executed first. This is followed by an acceptance test of the result it calculates. If the test is passed, then the result is accepted and passed on to subsequent parts of the system. If it fails, any side effects of the first are reset and the second section, called the first alternative, is executed. This too is followed by an acceptance test and is treated as in the first case. A second, third or even more alternatives can be provided if desired.

Conditions: Because of the possibility of having to undo the effects of program sections, the technique can only be used where the side effects can indeed be reversed. Special hardware may be necessary in some cases.

Major Advantages: Provides a form of diversity at the program, procedure or module level.

Problems or Disadvantages: It is not always easy to find reliable acceptance tests. If there is a failure, the temporal behaviour of the program changes quite drastically, which can be a problem in real-time situations. Undoing side effects of rejected sections may require special hardware. Reliability enhancement ultimately depends on both the acceptance test and the effective diversity between the sections.

Results: Increased run-time program reliability.

Related Methods: Closely related to the parallel diversity of multi-version software.

Assessment: Has been shown to be effective in situations without strict temporal constraints. May not be cost effective if a 'parallel' architecture is needed for other reasons, as then the parallel system could run a diverse single section program.

Tools: None known.

References:

1. *System Structure for Software Fault Tolerance.* B. Randell, IEEE Trans Software Engineering, Vol. SE–1, No. 2, 1975.

2. *Fault Tolerance, Principles and Practice.* (Vol. 3 of *Dependable Computing and Fault Tolerant Systems*), T. Anderson and P.A. Lee, Springer Verlag, ISBN 3-211-82077-9, 1988.

6.8   Redundancy

Aim: To detect and mask failures, to recover from failures.

Description: Makes use of more technical means than is necessary for the execution of a specific task. For example, the use of several channels, or several processors for performing the same task in parallel. Redundancy in its simplest form could be a duplication of equipment.

Conditions: An N out of M vote has to be performed to compare the results of the individual channel.

Major Advantages: Capable of detecting and masking random failures or discrepancies caused by poor modification or maintenance.

Problems or Disadvantages: There is a multiplication of hardware costs, and the need for a voter. There may be difficulties in comparing results if the channels are asynchronous, or difficulties in maintaining synchronization. This offers no cure against failures common to all channels. In particular there is no defence against faults in common software. Limited to hardware.

Results: Improved system performance.

Related Methods: Redundancy is fundamental to many other containment methods.

Assessment: Indispensable for safety-related hardware.

Tools: None known.

References:

1. *The Theory and Practice of Reliable System Design.* D.P. Sieworek and R.S. Schwarz, Digital Press.

2. *The Evolution of Fault-Tolerant Computing.* (Vol. 1 of *Dependable Computing and Fault Tolerant Systems*), Edited by A. Avižienis, H. Kopetz, and J.C. Laprie, Springer-Verlag, ISBN 3-211-81941-X, 1987.

3. *Fault Tolerance, Principles and Practice.* (Vol. 3 of *Dependable Computing and Fault Tolerant Systems*), T. Anderson and P.A. Lee, Springer Verlag, ISBN 3-211-82077-9, 1988.

4. *Fault Tolerant Computing Theory and Techniques.* Vol. 1 and 2, D.C. Pradhan (Ed.), Prentice-Hall, 1986.

6.8.1 N out of M Vote

Aim: To reduce the frequency and duration of system failure. To allow continued operation during test and repair.

Description: M channels are configured such that N outputs are required for action at the system level. Voting can take place using hardware or by operator interaction. The choice of N and M will depend on the system failure states that are of concern, and on the repair procedures adopted. Take a simple application where the computer system must trip the plant if certain dangerous conditions are detected. A number of different voting schemes could be used:–

1 out of 2 – one or both channels can require a system trip. Both channels must fail to prevent a correct response to a genuine demand. Either channel can cause a spurious trip when there is no demand.

2 out of 3 – two or more channels must request a trip. Two channels must fail to prevent a correct response to a genuine demand. Two channels must fail to cause a spurious trip when there is no demand.

The choice will depend on the reliability, availability and safety requirements of the plant. Consideration must also be given to the performance of the voting system when one of the channels is under test or repair. The channel can either be set to the trip condition or a no-trip condition (an operational bypass). The approach chosen will influence the choice of voting logic. For example a 2 out of 4 system with bypass, or a 3 out of 4 system with trip both reduce to a 2 out of 3 system when one channel is under repair.

Major Advantages:

- Reduces the frequency of all types of system failure that result from hardware failure at the component level;

both failure to act on demand and spurious system operation can be controlled.

- Allows test and repair without shutdown.
- Comparison of outputs enables the early detection of faults causing inaccuracy.

Problems or Disadvantages:

- Cost and resources necessary to provide and maintain M channels.
- Benefits limited by common cause failures.
- Many design errors in hardware and software are common cause, and would not be detected by the slight variation in channel inputs.

Results: Improves system performance.

Related Methods: A form of redundancy. Diverse software or hardware can be employed in the channels to enhance the probability of detecting and masking design errors.

Assessment: Essential for systems where any break in service has serious consequences. Critically dependent on the voter.

Tools: None identified.

References:

1. *The Theory and Practice of Reliable System Design.* D.P. Sieworek and R.S. Schwarz, Digital Press.

2. *Fault Tolerance, Principles and Practice.* T. Anderson and P.A. Lee, Prentice Hall, 1981.

3. *Fault Tolerant Computing Theory and Techniques.* Vol. 1 and 2, D.C. Pradhan (Ed.), Prentice-Hall, 1986.

4. *VOTRICS: Voting triple modular redundant computing system.* Proc. FTCS 16, Vienna, IEEE Computer Society Press nr. 703, 1986.

6.8.2   *N* out of *M* Vote, Adaptive Voting

Aim: To avoid that, in voting systems, fault masking ability deteriorates as more copies fail (i.e. faulty modules outvote the good modules).

Description: Adaptive voting is a technique in which the number of units taken into account in the vote can vary during the life of the system. The following techniques can be classified as adaptive voting techniques.Consider a voting system with *M* units (*M* Modular Redundancy):

Hybrid Redundancy – initially, voting is done between *M* units (for TMR: Triple Modular Redundancy, $M = 3$) and a further *S* units are considered as spare units. When a fault occurs in the TMR, the faulty unit is replaced by a spare unit.

Self Purging Redundancy – initially, the *M* units are taken into account in the vote (an *M* modular redundant system). After a fault occurs, the system reconfigures to an $M - 1$ redundant system, then to $M - 2 \ldots$ etc.

TMR simplex – initially, voting is done between 3 units and there is no spare unit. After a fault, only one of the non-faulty units goes on working as a simplex system.

Conditions: This technique requires the design of 'threshold' voters.

Major Advantages: Provides highly reliable architectures with efficient abilities of fault detection.

Problems or Disadvantages: It may be expensive. The design of the voter can be complex and so the voter may not be reliable enough.

Related Methods: Must be compared to pure voting schemes and detection/replacement schemes.

Assessment: Very valuable technique. Most useful in high availability systems where servicing is difficult or impossible. Critically dependent on the reliability of the voter.

References:

1. *The Theory and Practice of Reliable System Design.* D.P. Sieworek and R.S. Schwarz, Digital Press.

2. *Fault Tolerance, Principles and Practice.* T. Anderson and P.A. Lee, Prentice Hall, 1981.

3. *Fault Tolerant Computing Theory and Techniques.* Vol. 1 and 2, D.C. Pradhan (Ed.), Prentice-Hall, 1986.

4. *A Highly Efficient Redundancy Scheme: Self Purging Redundancy.* IEEE Transaction on Computers C-25, June 1976 .

5. *Reliability Estimation Procedures and CARE: the Computer Aided Reliability Estimation Program.* JPL Quarterly Tech. Review 1, October 1971.

6.9   Re-Try Fault Recovery

Aim: To attempt functional recovery from a detected fault condition by re-try mechanisms.

Description: In the event of a detected fault or error condition, attempts are made to recover the situation by re-executing the same code. Recovery by re-try can be as complete as a re-boot and a re-start procedure or a small re-scheduling and re-starting task, after a software time-out or a task watchdog action. Re-try techniques are commonly used in communication fault or error recovery. In this instance re-try conditions may be flagged from a communication protocol error (check sum etc.) or from a communication acknowledgement response time-out.

Conditions: The time-criticality of the re-tried software must be greater than the time required for a re-try attempt to succeed. Re-try may result in the loss of real-time data and therefore full re-initialization may be needed. Re-tries should be limited in number.

Major Advantages: Usually relatively inexpensive to implement. Under normal, correct, operating conditions minimal software overhead incurred.

Problems or Disadvantages: Most real-time systems cannot accommodate the loss of data and the time delay of automatic re-boot. Therefore they are mainly used in handling communication error recovery.

Results: Relatively straightforward recovery technique.

Related Methods: Normally uses either software time-out checks or watchdog timers to trigger re-try attempts.

Assessment: Should be used with care and always with full consideration of the effect on time-critical events, and the effect of lost data during re-boot.

Tools: None identified.

References:

1. *The Theory and Practice of Reliable System Design.* D.P. Sieworek and R.S. Schwarz, Digital Press.

6.10   Return to Manual Operation

Aim: To provide the operator or supervisor with the information and the means to perform the function of the failed automatic control system.

Description: It is applicable to a continuous process with human supervision and/or partial operation. The general procedure is to:-

- alarm the operator,
- transfer the failed system from automatic to manual operation,
- allow manual process control by the operator.

The operator would perform the same or possibly a reduced set of the functions performed by the automatic system. Manual operation should be described in the operating procedures and has to be approved by the safety authorities. In general, manual operation can only be permitted for a given period of time before the whole system has to be shut down.

Conditions:

- The functions must be safely manageable by the operator with the information available in the control room.
- The operator must be trained to follow the manual procedure.
- The operation must be authorized by the safety authorities.

Major Advantages: Increases the system availability, permits less redundancy in the automated controls and provides a diverse means of control.

Problems or Disadvantages: Requires continuous operator availability and specific training. Operators can be dangerously overloaded if too many systems fail at once.

Results: Higher system availability or reduced automation system complexity.

Related Methods: Related to redundancy and diversity.

Assessment: Useful provided it is used with care.

Tools: Computerized operator procedures.

References: None known.

6.11   Vital Coded Processor

Aim: To be fail-safe against computer processing faults in the software development environment and the computer hardware.

Description: The basic concept is to perform all computing operations on redundantly encoded data items so that processing errors can be detected as incorrectly encoded outputs to very high probability.

A vital coded processor consists of three components:–

- Vital input coders and output decoders.
- A standard microprocessor.
- A fail-safe dynamic controller which compares the 'signatures' of the output data with predetermined values and disables all output interfaces when an error is detected.

In this system, every input data item is scrambled to produce an associated signature. In the microprocessor these data items are processed by coded operators (typically implemented as subroutines). These coded operators are relatively simple functions (integer operators, boolean operators, conditional branches etc) which produce coded results. When a processing error occurs, it will affect the resultant data and signature. The error is then propagated to the output variable signatures by the subsequent coded operators. The final decoded signature is compared with a predetermined value. While it is possible that the predetermined signature is incorrect, it is very improbable that the *same* wrong value will be produced, so a fail-safe action will still be taken.

Vital coded processors have been used to implement the automatic train protection on Paris's express regional network. It is also used on the VAL of Chicago Airport and on the fully automatic metro Lyon D line.

Conditions: The concept must be an integral part of the overall system design.

Major Advantages:

- Guards random hardware failures.
- Provided the output signatures are calculated independently, it guards against systematic errors in the supporting hardware and software to very high probability. The actual probability of undetected error depends on the number of bits in the signature.
- System safety not affected by the microprocessor technology and the software support tools used to implement the application code. It is possible to use standard compilers, assemblers, linkers, real-time kernels without the need to validate them.
- The coded operators represent a 'virtual machine' which can be easily transported to a new processor technology.

Problems or Disadvantages:

- High processing overheads (possibly 100 times greater than normal).
- Decreased intelligibility of the source programs.
- Increased costs associated with the input encoding and output decoding.
- The coded operators are not suited for computationally complex applications.
- Determination of the expected signature is only partially automated. Some manual effort is required.

Limitations: Does not guard against faults in the software specification or software design.

Results: Provides a processing system that, to very high reliability, either executes its application program correctly or fails to a safe state.

Related Methods: A form of fail-safe design. Directly derived from the encoding techniques used on digital data transmission links. Related to failure tracking and dynamic logic.

Assessment: A novel form of fail-safe processing that overcomes most of the insecurities associated with microprocessor-based technology. Should be considered for use on relatively simple applications that have a safe state. Especially useful if a non-redundant system design is being considered, although it could be employed as a 'fail-stop' component of a redundant system.

Tools: A tool exists to compute signatures (MATRA Transport), but this is not generally available.

References:

- *Le Processeur Sécuritaire Codé*, P. Forin, Colloque AF-CET/PRDTT, Paris, Jan 1989.

Appendix

EWICS TC7 Contributors and their Companies

Barnes, M.	AEA Technology, Culcheth	UK
Bishop, P.G.	National Power (NP TEC)	UK
Bjarland, B.	Technical Research Centre (VTT)	Finland
Bloomfield, R.E.	Adelard	UK
Bologna, S.	ENEA CRE Cassaccia	Italy
Bove, R.	ENEA CRE Cassaccia	Italy
Covington, G.C. III	CAI	FRG
Dahll, G.	Institutt for Energiteknikk	Norway
Daniel, P.	Plessy Crypto	UK
Daniels, B.K.	National Centre for Information Technology	UK
Darbois, C.	Électricité de France	France
Ehrenberger, W.	Fachhochschule Fulda and GRS	FRG
Fergus, E.	Liverpool Data Research Associates	UK
Froome, P.D.K.	Adelard	UK
Galivel, C.	RATP	France
Gayen, J.T.	Institut für Verkehr, Eisenbahnswesen und Verkehrsicherung	FRG
Genser, R.	Bundesministerium für Verkehr Kovpol	Austria
Goring, C.J.	Triconex	UK
Gorski, J.	Institut for Informatics	Poland

Humphreys, R.	Rolls Royce and Associates	UK
Johnson, E.	ICI plc	UK
Kammerer, U.	Rheinisch-Westfälischer TÜV	FRG
Kanoun, K.	LAAS CNRS	France
Kirk, G.	Gordon Kirk Ltd	UK
Koornneef, F.	Delft University	The Netherlands
Krebs, H.	TÜV Rheinland	FRG
Krotoff, H.	Merlin Gerin	France
Lindeberg, J.F.	SINTEF ELAB	Norway
Lindskov Hansen, S.	Electronik Centralen	Denmark
Monaco, F.A.	Ansaldo S.Ṕ.A.	Italy
Nordland, O.	MBB ERNO Raumfahrt-technik	FRG
Ould, M.A.	Praxis Systems plc	UK
Pasquini, A.	ENEA-DISP	Italy
Pilaud, E.	Merlin Gerin	France
Poujol, A.	Comissariat à l'Energie Atomique DRNR-SPCI	France
Quirk, W.J.	AEA Technology, Harwell	UK
Rata, J.M.A.	Électricité de France	France
Redmill, F.J.	British Telecom International	UK
Rzehak, H.	University of the Federal Armed Forces	FRG
Schlesinger, J.	Softcraft AG	Switzerland
Schoitsch, E.	Austrian Research Centre Siebersdorf	Austria
Sintonen, L.	Tampere University of Technology	Finland
Smith, I.C.	AEA Technology, Winfrith	UK
Voges, U.	Kernforschungszentrum, Karlsruhe	FRG
Vogler, J.D.	Merlin Gerin	France
Wimmer,W.	ASEA Brown Boveri AG	Switzerland
Zalewski, J.	Association of Polish Engineers	Poland

Index